The Origin and Location of Our Universe

(We didn't begin with THE Big Bang...)

Joseph P. Zwack

**THE ORIGIN AND LOCATION OF OUR UNIVERSE
(We didn't begin with THE Big Bang)**

Copyright © 2007 by Joseph P. Zwack.

**Joseph P. Zwack
Cosmic Press
Third Floor** cosmic press
**100 Main Street
Dubuque, Iowa 52001**

All rights reserved.

For further information, including inquiries regarding discounts for bulk purchases, contact:

Cosmic Press, Publisher,
Third Floor, 100 Main Street
Dubuque, Iowa 52001

563-556-3407
Cosmicpress.net

Printed in the United States of America
First Edition printed in 2008
Graphics and design by Dzine Wise

ISBN 978-0-9801734-0-6

Dedication

For my wife, Paula.

And the kids, Laura, Joey and Sharon.

And the grandkids,
Andrew, Emma, Benjamin, Elizabeth, Claire, and Grace.

Contents

1. Introduction 7
2. The Official Position 12
3. The Simplest Solution: Cosmic Unity Theory 22
4. Problems With Big Bang No Problem For Cosmic Unity 46
5. Big Bang Evidence Proves Cosmic Unity Theory 74
6. What Kind Of SUSE Brought Us Into Being? 89
7. More About the Physical Connection Between Our Universe and the Surrounding Universe, and Random Thoughts 94
   Notes 114
   Index 124

1
Introduction

The age-old question that has been asked by many people in many ways remains unanswered: *What was the origin of our universe?*

The more I thought about the Big Question over the years the more I was surprised about something. It began to dawn on me that one of the simplest, most obvious and logical possibilities for the origin of our universe had never been seriously addressed so far as I could see. At least I've never seen it discussed by any of the popular authors.

I gradually became convinced that cosmologists and others were reaching for concepts of our origin that might be far too exotic, speculative and theoretical while at the same time overlooking the most evident. I found that I could reduce my general concept of the "most evident" theory to just a few sentences. When I jotted them down, though, it seemed best to give some context for them. In trying to do that, however, I felt I should at least point out the case for—or give the main proofs for—the major competing theory for the origin of our universe. That, of course, is the Big Bang Theory (BBT). But to put BBT into perspective I concluded that it was only fair to give the primary reasons why I and certain others rejected major parts and subparts of the theory. But a discussion of the weaknesses of BBT naturally led me to an explanation of how I felt the various weaknesses—and in fact impossibilities—of BBT were solved with my theory.

When I began setting my thoughts to writing it was my intent to end the discussion after making the main points that formed the basis for my concept of the origin of our universe. I would not venture into speculation about what lies beyond what I considered to be my answer to the Big Question itself. My plan would be to open the door to entry into my theory of the origin of our universe and let others take it from there. I only partly succeeded on that score. As you shall see, I couldn't help offering some speculation about what lies inside the next larger reality of the total universe. I ask the reader to please not let my speculation about things beyond the basic premises of my theory taint the basic theory itself.

I also soon found that a discussion of the origin of our universe led directly to an examination of the differing structure of our universe under BBT and my theory. Such differences are literally worlds apart.

In any case, my simple, several-sentence hypothesis has expanded beyond the original intent.

As I'm clear to point out from time to time in the following pages, my degrees are far removed from cosmology. At several points in setting my thoughts to paper (computer memory today) I contemplated getting mathematicians or physicists involved in respect to certain areas that arise in this discussion, but decided against it. My purpose in writing this is not to try to offer a mathematically-annotated scientific reference manual. I'm neither a mathematician nor physicist, and I won't try to appear to be such.

In this regard I have to admit that I was astonished at the total range of pure speculation in cosmology that is presented to readers by scientists of various stripe—much of which speculation seems to be based on little or no observational or foundational proof. The unrestrained guesswork presented by many of them makes my hypothesis for the origin of our universe seem tame, simple and pedestrian indeed. But the very simplicity of a theory embracing Cosmic Unity is, I believe, one of its greatest strengths.

I'm aware, of course, that non-professionals are often looked at with condescension and even some hostility when venturing into the world of cosmology. Stephen Hawking (who should not be counted among the condescending group) has actually bemoaned the fact that, because of the technicalities involved, it is becoming almost impossible anymore for non-professionals to discuss matters of cosmology; and that philosophers, generalists, et al. have for the most part been almost forced to surrender the field of cosmology to technicians. Physicist Lee Smolin has earned the wrath of the cosmological community by suggesting (among other things) that it is a mistake to virtually banish philosophy in broader discussions relating to the big ideas of the universe.

Nothing highlighted for me the hostility shown to the unwashed masses more than a television presentation of cosmology given by a panel of well-known cosmologists and space-science authors a short time ago. In the audience was a fairly large coterie of professors, students and scientists. The program itself was very good, but the question/answer period near its end was a huge disappointment. Assorted members of the audience seemed much more interested in impressing the other attendees and the television audience with how bright they were than asking meaningful questions. One obviously intelligent and well-spoken woman, who admitted she was not a scientist, complemented the panelists on how smart they were, and asked a general question about how the panelists felt personally about knowing so much more relating to our universe and its features than the general public had any sense of. As the television camera panned the audience, the smirks and guffaws at such an amateur's question were plain to see. Actually, though, a couple panelists did quite well in answering the woman's question.

A professor who was a member of the audience, however, would have none of it. When called on for his "question" he made a little speech about how fast our sun converts hydrogen to helium—which had nothing to do with any type of question for the panel—and then turned to the woman who had had the temerity to attend the lecture, which he obviously felt was intended for professionals such as he. He essentially told her that her type of person should simply content herself with astrology, looking each day at the newspaper to see what was in store for her astrological sign for the day, and live her life accordingly. Basically, she should leave the important questions to people such as he.

The shocked woman at that point was brave enough to speak up and say that that "was the most condescending thing I have ever heard in my entire life." As she was in the process of saying "just because I can't do the math" and therefore shouldn't be shut out of the conversation the panel moderator switched to another question. It was truly an embarrassing scene, even watching it on television. You can be sure, though, that the professor's comments had their effect: there were no other televised questions by non-professionals from the floor that day.

Even with that in mind, though, I'm going to say what I have to say.

•••••

We have to define what we mean by "our universe." When I discuss "our universe" during the course of this writing what I'm talking about, unless otherwise specified, is our commonly-accepted concept of our universe. Right now, most people believe "that" universe to be about 14 billion or so years old. It is that universe which is believed by many people to have begun with the Big Bang, and it is that universe whose beginnings and existence I shall discuss in the pages to come.

I include within the discussion of our universe that part of it that is thought to have begun with the Big Bang but lies beyond our visual horizon (because its light has not had time enough to reach us yet). Scientists presently believe that the visible horizon of our universe extends out about 14 billion light years, but that the distance to the farthest reaches of our Big Bang-created universe is much more than that. I use this wider definition of "our universe" mindful that it varies from the definitions of certain scientists. Some cosmologists only speak of the *observable* universe as within their working definition of "our universe." For example, Tufts Institute Director of Cosmology Alex Vilenkin notes, "Cosmologists...focus mostly on the observable part of the universe, leaving it to philosophers and theologians to argue about what lies beyond."[1] He does note, though, that it would indeed be a great disappointment if in fact the world did end at the horizon.

Wikipedia notes that "Both popular and professional research articles in cosmology often use the term "Universe" to refer to the *observable universe*. The reason for this usage is that only observable phenomena are scientifically relevant. Since unobservable phenomena have no perceptible effects, physicists argue that they causally do not exist." Throughout this book the terms "observable universe" and "visible universe" shall be used interchangeably.

California Institute of Technology professor Kip Thorne believes that the universe is "a region of space that is disconnected from all other regions of space, much as an island is disconnected from all other pieces of land."[2]

As you shall see, I believe just the opposite. I conceive *our universe* to be an integral part of a larger, total universe of unified, connected spacetime, as shall be discussed in the pages to follow. I do not intend to speculate much about the beginnings of the *total* universe from which *our* universe originated. Let's take this one universe, or part of a universe, at a time, starting with the origin of "our" universe.

Please do not take my criticism of various aspects of BBT as a gratuitous assault on its adherents. Many believers in BBT have in good faith invested a large part of their careers in teaching the theory or otherwise accommodating their professional lives to it, and the fact that BBT is probably wrong on various levels doesn't make its disciples any less professional, dedicated or honest. Much of the published specialized technical language one encounters with astrophysics, and certainly mathematical formulae relating to it, leaves me thinking that possibly three of my four years of Latin classes in school might have been better spent in some more math and science courses. I have utmost respect for the knowledge of the scientists with whom I take issue in the pages to come.

As the "establishment" position on the origin of our universe, though, the Standard Model BBT is the natural foil one has to use in order to show that an alternative theory is the more reasonable when measured against that most commonly accepted explanation. In fact the work of many BBT believers has actually produced the very evidence that in my opinion helps prove the true origin of our universe—and it's not "the" Big Bang.

When in my title to this book I refer to the "location" of our universe I of course give something away about my basic theory itself. Something cannot have a "location" without reference to other things. Thus my title betrays the fact that I do not think that what we consider to be *our universe* (both the observable and unobservable parts of it) is in fact the full extent of the total universe. You may be surprised to hear where I think we are, and how simple it all is.

My theory does not involve other dimensions, and it also certainly doesn't conform to any of the multiverse theories, as we shall discuss as we move along.

•••••

This is a very short book, and without some of the speculation in which I engage in the last chapters it could have been shorter still. I may not be able to prove to you today that my theory for the origin of our universe is right. I'm only trying to show that it's much more consistent, simple and logical than BBT or the other theories yet advanced.

2
The Official Position

Until recently the two main modern-day contenders as our closest answers to the origin-of-universe question have been the Steady-State (SST) theory in its various formulations and the hot Big Bang Theory (BBT) in its several formulations. SST is based on the concept of an ongoing, continuous creation of matter in a probably eternal universe; and BBT is based on a belief that an almost infinitesimally small, dense pinpoint of nothingness suddenly expanded (the "Bang") billions of years ago and stretched and evolved into everything that now comprises our universe.

Both theories have had detractors and have been beset with issues that couldn't be resolved. The Big Bang, though, has firmly moved into the most favored position as presently espoused by the great majority of the scientific community today. Some say that Big Bang Theory is really a theory for the evolution, or development, of the universe after the Big Bang, as opposed to a theory for the *origin* of the universe. I happen to agree that it fails as a theory for the origin of our universe, but the general public and most BBT devotees don't make the suggested distinction. Most refer to the Big Bang as the origin—even though unexplained—of our universe. *Wikipedia*, for example, states:

> The generally accepted scientific theory which describes the origin and evolution of the Universe is Big Bang cosmology, which describes the expansion of space from an extremely hot and dense state of unknown characteristics.

So throughout this book we'll refer to BBT as a theory for the origin of our universe.

SST has fallen into general disrepute in recent years (although there are various attempts to resurrect the theory). There are two primary reasons for the downfall of SST: The main one was the discovery of an all-encompassing background radiation that has been found to exist throughout our known universe. It's believed that such cosmic background radiation is the result of the Big Bang that brought our universe into existence some 14 or so billion years ago. That was bad news for SST. Second, it has also been pointed out that if our universe had always been in a steady state the farthest galaxies that we can now view with our best telescopes should look no different than the galaxies located much closer to us. What we actually observe, though, is that the most distant galaxies are possessed of irregular shapes and a higher population of very luminous, short-lived stars, etc. In other words, the distant galaxies are in fact much more primitive—thus thought to be *older*—than those close to us.

So BBT pretty much reigns supreme within the scientific community as the Twenty-First Century begins. But even the most die-hard Big Bangers usually admit there are some serious difficulties and some big unexplained questions connected with the theory.

Big Bang Theory is more a class of cosmological models than a single theory, but we'll refer to it throughout as a theory, or simply as BBT. BBT encompasses about a half dozen premises, several of which have off-shoots commonly recognized as a part of BBT by the scientific community, even if not formally embraced within the so-called Standard Model itself. In basic terms the main propositions of the Standard Big Bang Model are as follows: (a) The fundamental laws of physics we see on earth apply throughout our entire observable universe (This is sometimes referred to as the universality of physical laws); (b) Our universe is expanding; (c) Our universe is evolving; (d) Our universe originated in a state of extremely high density and heat (a singularity), which began all space and time; (e) Our universe today is properly portrayed by Einstein's concept of general relativity; (f) Our universe appears to be the same in all directions (is isotropic) and from every location (is homogeneous)[1] (These two concepts are sometimes referred to as the cosmological principle). As mentioned, built into almost every one of the foregoing propositions are various supporting concepts and conclusions.

When I talk about Big Bang Theory what I'm generally referring to, then, is a model that describes our universe that is thought to have begun about 14 billion years ago and is developing in accordance with the several basic propositions listed above.

The alternate theory for the origin of our universe that I shall set forth in the following pages is actually in either total or partial agreement with each of the above-specified basic propositions of BBT, but I do have very strong disagreement with portions of a several of them, and I think that various assumptions and conclusions drawn from some of them are fundamentally wrong. My disagreement on those several points in turn leads to entirely different conclusions from the Big Bangers concerning the beginning of our universe, its present and past status, and its future.

Let's begin. Whereas I agree that our universe began with a hot big bang, what I feel is one unacceptable conclusion as presented to us by the scientists is that BBT stands for a position that everything that is or ever will be in our universe came from *nothing*. BBT is considered to have begun everything. Let me repeat that. Every *thing* on planet Earth, every *thing* in our solar system, every molecule in our one-hundred billion-plus-star Milky Way galaxy, and every atom in every piece of matter in the one-hundred billion-plus galaxies of our universe is believed by the scientists to have begun with, and come from, the Big Bang's pinpoint of nothingness that had no beginning in time. (The actual size of the "thing" that suddenly sprang to life is usually claimed to have been a pinpoint so tiny as to be immeasurable.)

On this particular BBT claim—that *nothing* existed before the Standard Model Big Bang—let there be no question by anyone. It's important to understand this position, as it illustrates as much as anything else just how doggedly invested many proponents are in certain illogical aspects of BBT. A sampling of a few of the most popular authors of the day as they comment on our all-from-nothing beginning:

Paul Davies, author, theoretical physicist and professor of natural philosophy, states the matter succinctly: "What happened before the big bang? The answer is: Nothing."

Professor Davies has written that,

> [T]he Big Bang was the abrupt creation of the Universe from literally nothing: no space, no time, no matter. This is a quite extraordinary conclusion to arrive at—a picture of the entire physical Universe simply popping into existence from nothing."[2]

The brilliant theoretical physicist Stephen Hawking states:

> As far as we are concerned, events before the big bang can have no consequences, so they should not form a part of a scientific model of the universe. We should therefore cut them out of the model and say that time had a beginning at the big bang."[3]

When asked what preceded the Big Bang, Nobel Prize winner physicist Leon Lederman responded as follows:

> Well, the first thing is there's no "before." Because time itself, as far as we understand time, was generated—and space—at the Big Bang.[4]

Janna Levin, author of the popular *How The Universe Got Its Spots*, gives the standard scientific view when she states that "the big bang is the creation of space itself, of time."[5] And if such a statement leaves a question in anyone's mind about how she *really* feels about things before the Big Bang, she follows up with this:

> In the beginning there was nothing. I mean zero, nada, zip. No space, no time, no matter. Nothing."[6]

In *Big Bang* Simon Singh claims that at the moment of creation the universe was in "an unphysical state," and he points out that many cosmologists argue that the question "What came before the Big Bang?" is an "invalid" question.[7] But, not to fear, cosmologists appear to have hit on ways to rid themselves of that nagging, "invalid" question. One thing they do is just define it out of existence. As Singh goes on to state:

> [I]f time was created during the Big Bang, then time did not exist before the Big Bang, and it is therefore impossible to use the phrase 'before the Big Bang' in any meaningful way.

Levin says essentially the same when she states that,

> There is no sense to the question: how long was it before the big bang happened? Time started with the big bang.[8]

The NASA "*Ask An Astrophy*sicist" internet website goes with the "not meaningful" nature of beginning-time questions:

> The Big Bang certainly suggests that time began at the first instant of the Big Bang, since before then, the universe was collapsed into a singularity....The singularity is the point at which time has no meaning.[9]

One is also, incidentally, confronted with this same type of circular argument and definition when one asks a BBT disciple what is outside the boundaries of our finite universe: ("If space has no edges or boundaries, the question has no meaning.")[10] Such self-fulfilling assumptions and definitions allow avoidance of various obvious vulnerabilities of BBT relating to time and space. Patently, if one starts a conversation with certain unassailable definitional positions, then *ipso facto* any question that challenges such positions is simply "invalid," "has no meaning," or is "impossible" to be answered "in any meaningful way." And as we all know, he who defines the terms controls the debate.

As we shall later discuss at some length, the reality is that some of the BBT positions and resultant assumptions *are* assailable.

Paul Davies (mentioned above) goes with the "meaningless" position relating to what he acknowledges as certain "awkward questions."

> [T]ime itself began with the big bang. This neatly disposes of the awkward question of what happened before the big bang. If there was no time before the big bang, then the question is meaningless. In the same way, speculation about what *caused* the big bang is also out of place because causes normally precede effects. If there was no time (or place) before the big bang for a causative agency to exist then we can attribute no *physical* cause to the big bang.[11]

Of course what Mr. Davies says is true if his foundational "ifs" are true. Obviously, if there was no time there was no time, and if there was no place there was no place. But CUT holds that there *was* a time and there *was* a place before the Big Bang.

Cornell University has a popular *Ask An Astronomer* website, which has answered hundreds of thousands of astronomy-related questions in the past 30 years. Its team gives the same basic answer to the question at hand as is taught in the various Astronomy departments around the country. The all-from-nothing concept is so entrenched that people are not even supposed to *speculate* what might have come before the Big Bang:

> [W]e cannot speculate on what was there before since by definition there was no before. You can think of it as if time started to exist at the time of the big bang, so there was nothing before that.[12]

Even pop physics agrees with the more serious kind. Jennifer Ouellette, in *The Physics of the Buffyverse*, walks down the same mysterious everything-from-nothing path when she tells us:

> Technically, our universe arose out of nothing.[13]

A May 22, 2001 article in the *New York Times* attempted to summarize the positions of leading scientists relating to the origin of our universe. After quoting from several scientists, including the well-known Drs. Guth and Linde, the author of the article, Mr. Overbye, set forth the Standard Model position as follows:

[M]ost cosmologists, including Dr. Guth and Dr. Linde, agree that the universe ultimately must come from somewhere, and that nothing is the leading candidate.[14]

One could go on and on with the same type of quotations, as there are a great many statements by various scientists to the same general effect. As Mr. Overbye accurately noted, though, "nothing" is held to be the leading candidate for the origin of our universe under BBT. And, once more, just to be sure about it: Who makes such claim? The answer, once more, as observed by Overbye: "most cosmologists."

Those few BBT believers who can't quite bring themselves to swallow the totality of the everything-from-nothing precept of their theory employ several fall-back positions in addition to the definitional finesse mentioned above. Avoidance is again the key. Some professors just say that such an "origin" question is "not a profitable thing to think about," and refuse to try to answer it.[15] Others state, with barely-concealed irritation, that such a question is irrelevant. And some simply steer clear of the question in a variety of other ways, including a position that we should simply put off answering such a searching question.

Some scientists go so far as to tell us that "logic" shouldn't be used in trying to explain the everything-from-nothing problem (and of course religion is not a part of the equation). By the way, many scientists are not at all happy with people who feel that the universe somehow began with a prime mover. Here's the way Neil DeGrasse Tyson puts it:

> Yet certain types of religious people tend to assert, with a tinge of smugness, that *something* must have started it all: a force greater than all others, a source from which everything issues. A prime mover.
>
> In the mind of such a person, that something is, of course, God.[16] [Emphasis Mr. Tyson's]

What, then, are the above commentator's alternative scientific and reasoned positions on the beginning of our universe? His two possibilities, stated on the same page as his above comments: Either (1) "...the universe was always there" or, (2): "...what if the universe, like its particles, just popped into existence from nothing?"

Take that, all ye who seek scientific analysis in contradistinction to the ignorance of "certain types" of other people.

The theory I shall present has no hat in the ring on the question of religion. Earlier pronouncements from the Vatican (Pope Pius XII in particular), indicated that a big bang of some kind would be consistent with the position of the church regarding the origin of our universe. No pope has spoken *ex cathedra* on the subject. Other religious leaders have expressed views that seem to favor a Big Bang beginning. The rationale behind any such statements might work in favor, and might work against, the theory I shall advance in this book.

Returning to the (unsmug) positions of Big Bangers regarding the universe's start from nothing: Some argue, in so many words, that "our desire to find out what came before the Big Bang" is a "common-sense prejudice that should be set aside" in favor of Big Bang/Big Bounce theories, etc. There is also the somewhat diversionary claim by a few who assert that the rules of physics, gravity, chemistry, etc. did not, for some unexplained reason, apply either (a) immediately before, or (b) immediately after, the Big Bang, so questions relating to our universe's origin are unanswerable. And another more recent approach by BBT apologists is to point out that, since scientists think our universe harks back to a minute peck of something in the quantum realm—voila—no further answer is required. The magic word has been invoked. They're off the hook. After all, most agree that no one can really be blamed for not knowing what went on in the weird *quantum* world, right?[17]

But some people feel that it is indeed profitable to think about the question that is perhaps the most important one we shall ever encounter. These people don't agree that it's a surrender to a "common-sense prejudice" to seek answer to the Big Question. Count me among those who think that the origin of our universe is one of the most significant, absorbing, exciting and ultimately profitable questions one can possibly consider.

•••••

How, procedurally speaking, do modern day scientists even arrive at what they believe is the physical point of our beginning-from-nothing? The *modus operandi* of today's scientists is to calculate the life of our universe from today backward to the point in time (actually, it should be described as non-time, because nothing, even time, is said to have existed when the Big Bang occurred) to what they believe is the aforementioned tiny beginning pinpoint of almost-nothingness. But when they get there the scientists don't tell us what it was that went BANG, except to call it a "singularity." They can't tell us what went bang because if they could, then some *thing* existed to explode before the Big Bang and before time is said to have begun under their theory. That, of course, would be anathema to the singularity/Big Bang as our universe's originating mechanism.

As observed earlier, some Big Bang proponents don't like to dwell on the everything-from-nothing dilemma facing their theory. They would much rather concentrate on what happened after the Big Bang. It's almost as if, because the beginning point is said to have been so extremely small, we can just overlook the troublesome fact that that little "it" had no beginning. (Our non-existent pre-universe was evidently only a *little* pregnant.) It's so much easier for scientists to skip discussing what suddenly went BANG, by moving instead to the mechanics of post-bang things that can be reduced to measurements, calculations and formulae, complicated and impressive though they may be to peers and especially to those of us outside the professional cosmological community.

A perfect example of how the little "details" of the thing that began our universe are overlooked in order to proceed to post-bang things is seen in the position of the well-known physicist Alan Guth. Cosmologist Vilenkin describes what he calls Guth's "beautiful idea" regarding the beginning of our universe and its inflationary expansion (more of which later) as follows:

> A huge expanding universe was produced from almost nothing. All that was needed was a microscopic chunk of repulsive gravity material. Guth admitted he did not know where the initial chunk came from, but that detail could be worked out later. "It's often said that you cannot get something from nothing," he said, "but the universe may be the ultimate free lunch."[18]

So, all we need is that little unknown detail. And if we get past that detail we have a free lunch. Stated otherwise, if we can avoid impossibilities everything is possible.

The result of the whole BBT approach in its various forms is that we're essentially asked to throw up our hands and accept the entire mysterious pre-and immediate post-Big Bang theory on faith, if you'll pardon the expression.

However, in the humble opinion of this writer, logic and consistency tell us that some *thing had* to give birth to our physical universe; and it certainly should not be considered a "prejudice" to seek the identity of that causative thing.

Just think about it: Why should we have to accept the fact that every *thing* else in the world has a logical explanation (whether or not we've worked out its details yet) except one of perhaps the two most important and all-encompassing questions we could ever pose?[19] Why is it that the most graphic, tangible, obvious thing any of us will ever encounter should be relegated to a degree of magic, inconsistency and illogic that would normally be scoffed at by the scientific community?

However, as some have pointed out, BBT as presently professed not only avoids directly facing up to the Big Question, it actually demands that we deny one of the most fundamental laws of physics. That, of course, is the law of conservation of matter (and energy) that every student has heard of—and from that law we know that one simply doesn't create something from nothing. *Ex nihilo nihil fit.* But BBT dictates that we believe that not just something came from nothing, but that *everything* came from nothing. Consider the logic of such a position and ask yourself: Why would a basic law of physics that always applies everywhere we have ever observed within our universe *not* apply when and where it should seem most applicable—which is to that very universe that houses and is founded on such law? I'm sorry, but—primitive and unsophisticated as this may sound—it's a physical *impossibility* that everything came from nothing.

I want to stress again that the professionals who duck questions or define various problematic BBT questions out of existence are not bad guys and gals. Most of them know much more about math and physics than I'll ever hope to know. But they're working within a certain orthodoxy, a certain culture of approach to our universe, that is very peer-oriented. It's professionally safer not to attack the conclusions of esteemed peers and friends. Even professional funding is often tied up in the attitudes of peers. If and when I make some statements or take some guesses that prove to be wrong I'll survive.

Not so, with some of them.

Back to the subject of this chapter—the official position—the Standard Model Big Bang Theory adopted by the scientific community.

Many of us used to believe (and many still do) in a sort of oscillating universe involving a cyclic Big Bang. That theory held that there would periodically be a Big Bang; the universe would then expand for 50 billion years or so, then contract for 50 billion years or so, followed by a Big Crunch, followed by another recurring Big Bang. And on and on and on. A good number of scientists now say there are problems with that theory, not the least of which is that they feel there isn't nearly enough (observable) matter in the universe (called *critical density*)[20] to cause the universe to gravitationally slow down, contract, compress into a Big Crunch and start the cycle over again. On the contrary, scientists were astonished not long ago to learn that the universe appears to be expanding at a rate that is actually *increasing*, not slowing. But even if the cyclical Big Bang holds true—because we find more mass in the form of dark matter, because we discover that there are more neutrinos than expected, or because of some other reason—our beginning question still remains: What was the *first* cause? What was the actual *origin* of our universe?

Many people are still waiting for the answer to that question. Under any theory.

3
The Simplest Solution: Cosmic Unity Theory

One thing I learned in the practice of law over a span of more than half my life is that the most reasonable explanation for a problem usually turns out to be the right explanation. Once in a great while a totally unexpected answer to a predicament suddenly pops up, but in almost all cases an impartial court and impartial jury will find that the most reasonable explanation is the right one. A corollary of the above rule is that the most reasonable explanation to a question is often the simplest one. Again, there are occasional exceptions to the corollary, but for the most part its best to stick with the basics. Even the most profound things many times turn on the simplest solutions. Perhaps the most beautiful and insightful formula of all time is the simple equation $E = MC^2$. So, simplicity and reasonableness will be our polestars throughout this discussion.

With these things in mind let's directly address the Big Question regarding the origin of our universe. We will find that we don't have to rely on far-fetched fine-tuning. We will find that the answer to our ancient question of causation could be the simplest possible answer of all. And it's tied in to some readily-observable things.

My suggestions relating to the origin of our universe are founded on just a few simple propositions. I'll list them and then later comment on each in order:

1. **Everything is or was a part of something larger. This is the underlying basis of what we shall refer to as our Cosmic Unity Theory (CUT).**

2. **Our universe is no exception to this rule, and thus our universe is a part of a large, total universe.**

3. **Since the above two statements are true *our* universe probably owes its existence to the occurrence of some event in the large, total, or parent[1], universe of which we are a part.**

4. **In view of the foregoing simple propositions, one need**

merely to decide the nature of the parent universe and decide what sort of event it was in the large, total, parent universe that brought our universe into being. CUT's candidate as the event that most probably caused our universe to come into existence is some kind of star explosion. Under CUT this event is referred to as a *surrounding universe star explosion* (SUSE)). Our universe is composed of the expanding and evolving remnants of such surrounding universe star explosion.

I'll discuss the nature of our surrounding parent universe as we proceed.

If the foregoing statements are correct and in fact our universe owes its existence to a SUSE we solve dozens of questions and problems that now confront the various existing Big Bang formulations.

Let's discuss in a little more detail the above four statements, starting with the comprehensive, cohesive theory of Cosmic Unity.

1. The Cosmic Unity Theory. Every thing we know of is or was a part of something else, which is to say that every thing is or was a part of something *larger*. Think about it. As a purely observational matter, and even intuitively, we can realize that all material being that we see, hear, smell, taste, touch—whether directly through our senses or through scientific measurements or calculation—is or was a part of something larger. This fact holds true for every item, object, part, collection, and/or physical *thing* of every kind—from the very smallest to the very largest, from quantum to astronomical. There are no known exceptions to this rule. In fact, I would recommend that this property of existence—that all things are or were a part of something larger—be added to the rather short list of true metaphysical "universals."

As we shall see in the following pages, many of the Big Bangers' basic beliefs about the development of our universe can be applied with full force to CUT, except that instead of tracing our universe's lineage back a troublesome, contradictory, pinpoint of nothingness that contained the entire mass of our universe and did not exist in time or space and had no beginning, under CUT we instead trace our lineage back to an explosion of a relatively large object that actually existed in time and space. Simple as that. After the explosion of our parent universe progenitor star CUT is quite consistent with many aspects of BBT. Consistent, that is, except for the theses and adjuncts of BBT that just don't make sense.

In point of fact, BBT about our universe's evolution works pretty well developmentally for everything following a big bang on up to the present. It just doesn't work for a beginning; it does have a few practical problems in the middle; and it doesn't comprehend the end of our universe.

•••••

Astronomers sometimes use smaller-to-larger graphics in the form of "tours" to illustrate just how immense our universe is, and the reader of these words may well have seen such smaller-to-larger graphics. For ease of illustration multiples of ten are sometimes used. For example, we start in a theoretical spaceship at a certain point on earth—say our front lawn—and then move out into space and look down on our front lawn from 10 meters, then 100 meters, then 1,000, then 10,000, etc. It doesn't take long before we're out of our solar system, our galaxy, our Local Group of galaxies, our Virgo Cluster of galaxies, and eventually out to the farthest points we can possibly see with telescopes optically or in other wavelengths. (As mentioned earlier, we believe this farthest *observable* distance from earth is about 14 billion light years or so.)[2]

Moving out into space in distance multiples of 10 ("orders of magnitude" of 10 or "powers" of 10) then, is one shorthand way of getting at least some perspective on the mind-boggling immensity of our observable universe. (Of course the farthest extent of our universe that started out long ago with a bang reaches out much beyond what we can observe today.)

Using the imaginary spaceship and graphics mentioned above we can then zoom back from space and magically shrink ourselves into a tiny spaceship traveling microscopically into smaller and smaller realms. If the beginning point of our journey inward happened to be, say, from our grass-covered front lawn we can then decrease the size of area we are examining from one of the individual pieces of grass on which our spaceship was sitting downward tenfold into the grass blade's cellular structure, then into its sub-cellular structure, its molecular structure, its atomic structure, its subatomic structure, etc. in decreasing magnitudes of ten until we reach attometers or smaller in size. Here we're entering the world of leptons, gluons and quarks of various color and flavor.

Depending on whose illustrations one uses, and which basic measurement of distance we use—centimeters, meters, inches, feet, miles, etc.—let's assume it might take about 50 powers of ten to rise from the smallest thing in our universe to the largest observable structure (keeping in mind that the actual size of our universe today has moved far past the observable 14 billion light year horizon in all directions). One good set of such illustrations I viewed involved some 43 such powers of 10 progressing from the smallest to the largest. (Note: It's said that there are actually about 61 orders of magnitude difference between our smallest Planck length of 10^{-33} centimeters and the farthest distance we can see (our universe's horizon) of about 10^{+28}). The particular illustrations on the tour I mentioned above didn't go further than 43. Why? Because the illustrators had reached the most distant objects we can observe in our universe, which, as mentioned, are some 14 billion light years away. (These illustrators used 14 billion light years as the extent of our universe, even though, as discussed above, I am including within my description of the actual size of our universe that part of it which is beyond our present visual horizon.) It's believed by scientists that some of the farthest objects in our universe formed only a few hundred thousand years after the Big Bang, which is only a couple seconds into the 24-hour time clock we sometimes use in illustrating the passage of time from the universe's beginning until now.

So let's consider this. Let's assume *for illustration purposes only* that, using multiples of 10, in point of fact it takes precisely 50 enlargements to stretch from the smallest structure in our universe up to the very largest observable object, to wit, the entirety of our observable universe that began with a big bang some 14 billion years ago. Again, we're going to assume for our discussion only that it's an actual, proven, undeniable fact that the order-of-magnitude figure is 50. No more, no less. Then, once we reach that farthest visible point of our universe we continue outward until we reach the farthest point to which our universe has expanded past our visible horizon.

When we reach that far final distance CUT's position is this: We will not be forced to stop our theoretical spaceship at such farthest multiple of distance comprising our universe. If we try to move past that far-flung location[3] we will not encounter (1) an impossibility in the nature of nonexistence, (2) an impossibility in the nature of the shape of space, or (3) newly-created-space that arguably might be needed to accommodate our presence. Rather we will simply move with no difficulty one or more enlargements out into the real, pre-existing spacetime of our surrounding parent universe.[4]

I believe that something that holds true on each occasion in our universe for decillions upon decillions of times in a row (Ie., that an object was a part of something larger) in all probability will not suddenly become false by adding one more number. If we were to assume that the farthest multiple of distance to which we could travel in our universe is, say, X, it's my belief that traveling out to X plus 1 would, so far as physical travel is concerned, be just the same as traveling out to X. Again, we have the understandable visual problem that we simply can't see past a certain distance with present-day instrumentation, plus the fact that light from outside our horizon hasn't had time to get here yet.[5] But my point is that there is no barrier that we reach at the farthest point in our universe, and there was none at the time light started our way billions of years ago—and we do not encounter, and would not have encountered, any kind of barrier past that point.

After reaching multiple number X we simply travel into, with no recognizable change in condition, the real, finite world of the surrounding parent universe. This is the concept of CUT, and it is, of course, very anti-BBT.

•••••

What has impressed me my entire adult life is the order and consistency of life in general and our universe in particular. To me our universe seems to be obviously not random. Everything, from smallest to larger to larger, appears to exist in a continual flow, one level supportive of the next. Greek philosophers, Sir Isaac Newton, Albert Einstein, Max Planck and Carl Sagan (among many thousands of others) have all taken cognizance of and commented on the obvious, persistent, and pervasive coherence and order of the universe. (The very order of the universe was one of St. Thomas Aquinas' proofs for the existence of a God. Cosmic Unity does not as a part of its theory have a position on the existence of God, but it should be kept in mind that many atheists believe in the order of the universe dehors a God.)

Even in the world of quantum particle physics one sees (not literally) the order, pattern and symmetry of the universe. Such order is acknowledged to be so dependable and consistent that physicists are now able to refer to a "Standard Model of Particle Physics" that includes all matter.[6]

It's true that one's definition of order might vary to some extent depending on who's doing the defining. Although varying disciplines may offer different particulars for their individual definitions of order, the general concept of symmetry and dependability would be inherent in each definition.

But, as always, some people can't see the universe for the stars when it comes to the order of the universe. One popular writer of the day argues that the concept of order in our universe shouldn't apply because of what he calls rogue asteroids and comets colliding, many other things in our universe crashing into each other, galaxies cannibalizing each other, black holes gobbling up stars and planets, and stars collapsing and exploding, etc. Even on earth, he points out, we have predation from wild animals, disease, nature. He believes that the order that earlier scientists were mistaken in believing in was masked by their inability to see the true world as we, with our sophisticated telescopes, now see it to be. What we now see, he believes, is his version of disorder.

Hogwash, I say. Such an anthropocentric view of the physical world overlooks the fact that all his examples of disorder occur as they do according to ordered rules of nature. Cosmic things crash into each other because of gravity or the laws of motion, not because of capricious whims or unknowable and disordered forces. On earth big animals eat little animals for a good and understandable reason, not because of "disorder." Death by disease doesn't occur by lottery. Disease has a scientific basis.

We humans may not always be personally happy about the immediate results of what happens in a particular instance in the world of nature, but such a personal viewpoint gives us no right to declare that our universe is thus "disordered." We humans may fear the crash of a comet, and we certainly hate and dread the results of cancer; but these phenomena occur because of the reliability and order of the rules of nature. We can only hope to cure cancer someday *because of* the order of our universe. As a matter of fact we could not exist in a world of chaos and disorder, and often what we deem to be chaos proceeds from certain rules of order. Physical actions and reactions occur for a reason. Ultimately, order rules. Always.

It's true that the known facts relating to the order of the universe have developed and increased as time has passed and as humans have been aided by evermore sophisticated instrumentation, but the belief in the order of the universe remains solid among the great thinkers of our time.

Some could argue that the concept of "entropy" works against Cosmic Unity. Entropy refers to the measure of disorder in a closed system, and holds that every *process* that takes place in such a system leads to more and more disorder. Disorder is always supposed to increase as processes take place. More than 200 years ago German physicist Hermann von Helmholtz stated that, because of entropy, which is founded on the Second Law of Thermodynamics, the processes in our universe would eventually lead to its "heat death" (referring to our universe's *lack* of heat after burning up all its fuel and using up its energy as the world becomes more and more complicated.)

I'm certainly not an expert on the subject, but I would respectfully disagree with Mr. Helmholtz and those who agree with him on this particular point. From what I can see, the concept of entropy is an often misapplied principle, and it should, by definition alone, not be determinative of the end of our universe—if for no other reason than that it can be pointed out that Helmholtz conceived of our universe as a *closed* system. Under CUT, of course, our universe is an *open* system that forms a part of a great, surrounding, cohesive, unified, total cosmos. Whether entropy will eventually have a debilitating effect on our parent surrounding cosmos countless of eons hence can be discussed some other time.

In the meantime, what we see in our universe is order built on unvarying order, and I leave it to others to explain why entropy isn't noticeably affecting such observed order. Some scientists claim that most of the entropy in our universe appears in the form of cosmic radiation that resulted from the Big Bang, and that very little entropy is observable to us outside of that phenomenon. On the other hand, astrophysicist David Layzer contends that entropy and order can both increase at the same time without violating the second law of thermodynamics. All that seems certain to me on the subject is that what we see in our universe is reliable and unvarying order—in our measurably historical past, now, and in the foreseeable future.

So, I take the order of the universe as one of the proofs for Cosmic Unity, although CUT is certainly not dependent on it. The argument can even be made that even disorder necessarily implies the existence of Cosmic Unity, because disorder, like order, has to have reference to other things, which is a larger reality. The concepts of order/disorder are systemically interwoven with acknowledgement such greater reality. I happen to think that that greater reality is order.

Webster's Encyclopedic Unabridged Dictionary gives the applicable definition of "order" as:

> a condition in which each thing is properly disposed of with reference to other things and to its purpose; methodical or harmonious arrangement.

I must say that it seems totally *inconsistent* to me that the unfailing system of ascending and descending order and patterns we always observe at all levels everywhere can be consciously denied as it applies to our universe simply because we on planet Earth cannot at this particular moment in time literally see the bigger picture. But those who would continue to adhere to BBT will have to insist that the usual principles of order and unity always observed otherwise are violated in the case of our universe itself. Under BBT our universe *cannot* be a part of something larger, because if that were so, BBT's singularity could not have started everything. Thus under BBT when we reach the boundaries[7] of our universe, CUT's suggested rule that everything is a part of something larger would have to be arbitrarily disposed of by Big Bangers. I say "arbitrarily" because there is no good reason for disposing of the rule.

Obviously, then, I can't agree with BBT's stop sign that seems to me to be capriciously posted at the physical limits of our universe. The very existence of our universe should automatically have reference to something else—to our next larger component of Cosmic Unity. We're perfectly aware that we humans cannot visualize our total cosmos with today's tools and the limits imposed by the speed of light. The total universe may in fact not only be more than actually meets the eye today (acknowledging that there was an original opaqueness to our universe), but also more than is found within our universe beyond our visible horizon. We should be open to the fact that the odds are extremely high that the usual order of things—including the fact that we are all a part of something larger—will apply regardless of humanity's present scientific inability to prove or disapprove such fact in the next hour or so. Reality doesn't stop at the horizon, and it doesn't even stop at the farthest stretch that our universe has reached since our universe's beginning big bang.

Everything is or was a part of something larger. It's a proposition that cannot be proven wrong.

2. Our universe is a part of a larger total universe. If one accepts the reasonability of the proposition that *everything* is a part of something larger, then of course "everything" includes our universe itself, and that part of the case is closed. My belief that our universe should be included within "everything" is at least partially based on a sort of reverse proof: It should be the burden of the skeptic to disprove that a principle that applies at all times and everywhere else should *not* apply to just one additional case (our universe). Especially is this true when non-application of the principle requires an unabashed belief in the goodly number of physical miracles (often euphemistically referred to as "fine-tuning") that are demanded to keep the alternative concept of BBT afloat. We will go into these contrivances is some detail in Chapter 4.

Precisely why should our universe under BBT be an exception to the general rule that everything is a part of something larger? Because it's big? Because we can't see past certain levels of our universe today? I don't see the magic in the number X (or 61, or whatever the actual order of magnitude is) that should stop application of the rule of Cosmic Unity that inheres in decillions x decillions x decillions of instances throughout our universe without one single known exception. As noted before, if something holds true in such uncountable instances in our universe I think it should be up to any critic of the rule to present hard proof to rebut the presumption that it will hold true one more time.

It's true that, for at least a couple reasons noted above, we can't see beyond about 14 billion light years distance. We make estimates and calculations about how far past the horizon our universe extends. But think of all the occasions down through the ages that we thought that the universe didn't exist farther than we could see or calculate at the time. It was just a blink of time ago that we were able to see and comprehend for the first time that there are island universes in the form of galaxies that exist entirely outside our own galaxy. Even the Great Debate of 1920 between Harlow Shapley of Mount Wilson and Heber D. Curtis of Lick Observatory didn't settle the question whether the "nebulae" that were observed through the large telescopes of the day were separate island universes or a part of our Milky Way. Despite the fact that Einstein and many of the other great minds of the day were in attendance, people left the meeting hall with the question still unresolved.

I don't know how long it will be before we will be able to see (1) *out* of our part of the total universe and into that part of it occupied by our surrounding parent universe, or able to see (2) the intrusion *into* our universe of something that was originally outside the confines of our universe. This may involve somehow solving the "visible horizon" problem in the sense that we can't see past things where insufficient time has elapsed for light to reach us traveling at 186,282.396 miles per second. (Actually, I'm confident that someday in some way such observations will come to be.)

CUT takes the very simple, logical step of acknowledging that our universe itself is not an exception to the reality that everything is a part of something else. Historically, our universe is a natural result of a very common occurrence in a unified cosmos. A belief in such Cosmic Unity should certainly seem much more reasonable than one that says that (1) the limit of "everything" just happens to coincide with what people can see with the instrumentation and can calculate with the computers used in the first part of the 21st century after Christ was born on the planet Earth, and that (2) people were wrong every single time they believed in such a coincidence down through history, but this time they're right.

Needless to say (then why am I saying it?), I disagree that humanity just happens to be facing such a scientific coincidence. Again.

⋯⋯

If Cosmic Unity is true, then what would be the structural connectivity between our universe and our next larger parent universe? Cosmic Unity deals with (among other things) both the *origin* and *structure* of our universe. I guess we have all probably seen illustrations representing various concepts of a multiverse. (CUT doesn't conform to the suggested multiverse theories, as we'll explain later.) The multiverse concepts usually embrace some type of bubbles or pear-shaped things adrift in, well, something or nothing. The theories often conceive of connections by way of passageways, tubes, umbilical cords, wormholes[8] or black hole relationships; and sometimes they have no physical connections at all. As we shall soon discuss at greater length, any such universes other than our own universe are conceived of by the theorists as having different sets of physical laws, thus causing the universes to remain forever incommunicado with each other (though some argue that there could be infinite possibilities built on infinite possibilities out there. I'll address this later.) But under the usual multiple universe/multiple-physical-laws-concept things are all quite complicated and more than a little discouraging as seen by the theorists.

Under CUT, though, the morphological connection between our universe and our parent is as simple as can be. *Our universe is a part of, and surrounded by, the finite, pre-existing space of a larger cosmos.* No passageways lie between the two; no tunnels, bubbles, foam, walls or trapdoors are involved; there is no necessity to travel through black holes to get from one to the other. *And* there is no difference in basic laws of physics. Further, we are not in a different dimension from our surrounding universe. In fact, as more and more time ticks by under a Cosmic Unity theory, the spread of our SUSE's expanding remnants in the form of our universe will become ever more diffuse and incorporated into the surrounding medium. It will become harder and harder to distinguish between what is "our universe" and what is or was our surrounding parent universe. Like a splash in a pond, the big waves near the splash reach out and become smaller waves, then ripples, then barely discernible undulations, then nothing that is perceptibly different from its surroundings. The original splash ultimately just dissipates and is incorporated into the pond itself.

If we look at supernovae in our universe we see that that is pretty much what always happens after the passage of many of years. The expanding remnants, in time, dissipate into the surrounding space. CUT holds that this is essentially the same thing that is happening following the birth of our universe. *Our universe can be considered the remnant blast field of an explosion that occurred billions of years ago in the surrounding parent universe.*

We will speculate some more in Chapter 7 about the physical connectivity between our universe and our parent universe.

3. Our universe owes its existence to the occurrence of an event in our parent universe. As before, if one accepts the proposition that everything is or was a part of something else, this statement Number 3 is merely an extension of that proposition. Things have consequences. In our universe, we see that each action has some type of reaction. Although Newton's Third Law of Motion may well be altered in certain cosmic instances I believe that his general principle would apply to the basic truth we're discussing here.

Here on earth if one chops into a tree with an axe, many smaller worlds are affected by the bite of the woodsman's axe. The same thing happens when we chew on a piece of food, when a firecracker goes off, when we shovel out a grave, and when a rock rolls down a hill. All actions have consequences at various levels of magnitude. Even on the larger scales of a celestial plane we have seen what happens when comets strike planets, and when planets, stars and galaxies collide. Sometimes new worlds are created and sometimes they're destroyed or changed forever.

It just happens that awhile back a very common event occurred—one that does not tax credibility or require speculation about exotic and abstract hypotheses that deny normal application of the rules of science—and our universe is the result of such common occurrence. A supernova occurs about 50 times per second in our visible universe. One wonders how often a supernova occurs in our immense surrounding *parent* universe? Consider all the collisions and mergers and implosions and explosions that trigger huge reactions in our universe every day—then just try to imagine how often such things happen in our fantastically-larger surrounding parent universe.

Perhaps the biggest obstacle to overcome in discussing a CUT is that most scientists haven't really thought about a finite, interactive, physically-*surrounding* universe. Not really. Some might politely nod their head in the general direction of what many call a megaverse, multiverse, pocket universe, endless universe, perpetual universe, bubble universe, etc. but that's not at all what we're talking about. In my opinion even those who do note the possibility of a larger cosmos don't conceive of the most reasonable path leading to and from our universe and a larger total universe, and certainly not of the type we're talking about. It seems to be almost a requirement for those who acknowledge the possibility of some sort of universe other than ours to effectively cut off discussion about a CUT-type reality by steering, probably unintentionally, the discussion down the wrong path. The apparent and oft-stated belief is that if a universe other than our universe exists, it possesses other laws of physics and, after all, how can one much discuss another universe if all laws of physics and connectivity between the universes are unknown and/or undecipherable? Result: End of meaningful analysis that might lead to a realization of what we're now discussing.

There isn't an actual analysis of anything so simple as a true physically-surrounding parent universe of the type contemplated by CUT. One with laws of nature *compatible* with ours as part of an existing, cohesive, total cosmos. The seeming obligatory group-speak that effectively forecloses intense examination of the subject is remarkably similar in language.

The following are just a few examples of how leading authors almost automatically tie together the possibility of worlds outside our 14-billion-year-old universe with the probability of *other sets of laws of physics* existing in a multiverse-type setting (with emphases supplied).

From astrophysicist Fred Adamses' *Our Living Multiverse*:

> The multiverse produces a wide variety of different universes *with different versions of the laws of physics*.[9]
>
> ....
>
> Moving vicariously beyond our universe we thus find ourselves within a vast assemblage of other universes, *each with its own laws of physics*.[10]

Author Michio Kaku of City University of New York states it simply:

> [T]here could be an infinite number of universes *each with a different law of physics*.[11]

Professor Michael Duff states:

> The other universes are parallel to ours and may be quite close to ours, but of which we'd never be aware. They may be completely different with *completely different laws of nature* operating.[12]

Wikipedia incorporates a difference in physics as a part of the very definition of the term multiverse. Multiverse:

> "assumes the existence of many universes with *different physical constants...*"
> *[/wiki/Multiverse_%28science%29]*

Simon Singh, in his 500-page *Big Bang* compendium, defines the multiverse as:

> An alternative model to the single universe, in which many different universes co-exist, *each accommodating a different set of physical laws and each isolated from all the others.*[13]

Alex Vilenkin, author, professor of physics, and director of the Tufts Institute of Cosmology, believes we live in one of an infinite number of self-contained island universes. We will never be able to interact with other another universe under his theory:

> [A] voyage to another island universe is impossible, even in principle.No matter how long we travel and how fast, we are forever confined to our own island universe.

He believes that the laws of physics, which he refers to as "constants of nature," would differ in each of his infinite number of island universes:

> Constants of nature that shape the character of our world *take different values in other island universes.*[14]

So also, in talking about the possibility of other universes, the very readable Brian Greene, in *The Elegant Universe*, states,

> The central observation is that whereas ....we noted that everything we know points to a consistent and uniform physics throughout our universe, *this may have no bearing on the physical attributes in these other universes* so long as they are separate from us, or at least so far away that their light has not had time to reach us. And so we can imagine that *physics varies from one universe to another*....If we let our imaginations run free even *the laws themselves can drastically differ from universe to universe.* The range of possibilities is endless.[15]

....
Across the entire multiverse, these features vary widely; their

properties *can* be different and *are* different in other universes.[16] *[Emphasis Mr. Greene's]*

Some samplings from Martin Rees's *Before the Beginning*:

> What's conventionally called 'the universe' could be just one member of an ensemble. Countless others may exist in which *the laws are different.*[17]
>
> ....
>
> Each universe starts with its own big bang, acquires a distinctive imprint (and *its individual physical laws*) as it cools, and traces out *its own cosmic cycle.*[18]
>
> ....
>
> Could there be other universes, perhaps *governed by different laws?*[19]
>
> ....
>
> The multiverse could encompass *all possible values of fundamental constants*, as well as universes that follow life cycles of very different durations....some may.... expand for more than 10 billion years; others may be 'stillborn' because they recollapse after a brief existence, or because the physical laws governing them aren't rich enough...In some there could be no gravity....In others, gravity could be so strong that it crushes anything large enough to evolve into a complex organization....Some could even have different dimensions from our own.[20]

Answering a question on the general subject Kings College Royal Society Professor Sir Martin Rees goes on to state:

> But perhaps what we've traditionally called our universe is just an atom in an ensemble — a multiverse punctuated by repeated big bangs, where the *underlying physical laws permit diversity* among the individual universes.[21]

Andrei Linde proposes an infinite collection of wormhole-connected "bubble" universes created by a type of quantum "foam." These universes cannot interact with each other, however, because of—you guessed it—differing laws of physics in each universe. In fact, Mr. Linde essentially says that the multiverse is by definition an assortment of other universes with differing laws of physics.[22]

Paul Steinhardt of Princeton states that the possibility of other universes would not be testable in any way. Why? "Because," he says, "the different universes would not be detectable by one another." Any talk about multiple universes, states Steinhardt, means you're not even talking about science anymore: "In my view, you're into metaphysics."[23]

Science journalist, author and lecturer Andrew Chaikin agrees that communication between other universes would be impossible. He states,

> But there's a problem. If these other universes exist, there's no way for us to detect them.[24]

Those who conceive of a multiverse as an infinite number of pocket universes sing the same song. Paul Davies states that when we go from one pocket universe to another "things would look very different," and that our universe would be "very atypical."[25] Davies conceives of the multiverse as made up of "entire other universes."[26] Such a view of the cosmos, of course, is a far cry from Cosmic Unity, based as CUT is on *one* entire, total, cohesive, physically-surrounding, probably-compatible universe.

And if the other "verses" appear as three-brane worlds, the position of the experts is the same. Braneworlds (brane is short for "membrane") are theoretical three-dimensional bubble-world spaces somehow existing in a higher-dimensional space. Lisa Randall states the belief as follows:

> Branes could have different dimensionality.
> ....
> New particles with which we will never directly interact might propagate on such other branes.
> ....
> The other branes will probably be nothing like our own.[27]

The foregoing quotations are just a sprinkling of the representative conclusions that if our universe is or could be a part of something larger, then such larger thing has to be a vast number (probably an infinite number) of exotic, mysterious multiverse members whose individual laws of physics contradict each other from one incomprehensible universe to the next incomprehensible universe.[28]

....To all of which I want to yell out: "But what if our universe is simply the remnants of a surrounding parent universe explosion, and from which we have inherited the *same* general physical laws!? *Not* (a) exotic other brands of physical laws in (b) other total universes....But the same general laws as our birth parent. *What if?*"

Actually, all CUT asks is that we extend one of the most basic propositions of the BBT Standard Model. As we noted in Chapter 2's "Official Position," it's a foundational concept of BBT that the basic laws of physics we see on earth apply throughout the observable universe. It's the so-called universality concept. All CUT does is note that such laws will apply farther than we can see today.

Let's give the concept of the universality of physical laws real application.

The basis for the conviction of most cosmologists that any other universes undoubtedly have other physical laws arises primarily from the notion that our early universe underwent a period of intense chaos, and that it would be extremely unlikely that another universe going though the same type of chaotic beginning would come out of such turmoil with the very same values for a cosmological constant, the mass of particles, and laws of physics as did our universe. They feel that the chances of such a coincidence would be almost nil, and the result is that our universe would be faced with an impossible task attempting to communicate with any such other universes that developed so differently, with vastly differing characteristics.

Cosmologist and author Paul Davies feels the same way, but speaks in terms of "symmetry-breaking." He believes that a universe that began with a hot big bang will combine with "symmetry-breaking"; and that in the absence of some grand unified theory, the "default assumption" is that the universe we observe is merely one fragment among a haphazard patchwork of universes with differing domains, properties and constants of nature.[29]

Such a dilemma (the inability to communicate with others because of vastly-differing laws of physics) confronting differing universes might well be true, and, if true, would present a rather discouraging eventuality.

But our Cosmic Unity Theory makes such concern irrelevant.

Under CUT we are *unified* with the large parent universe that physically surrounds us. Our universe consists of the evolving remnants of a progenitor star or stars that existed in our very same total universe. *Same total* being the operative words. *Our universe will no more be alien to the surrounding total universe than our solar system is alien to the supernova that caused us to evolve from its remnants.* Our universe's basic physical laws will not be incompatible with the rest of our family, the result of which is that we may very well be able to interact with entities within (and other possible other progeny of) our surrounding parent universe. Although for identification purposes I have often been, and will be, using such phrases as "our universe" on the one hand, and "parent universe," "next larger universe," "surrounding universe," "total universe" and "total cosmos," etc. on the other, the fact of the matter is that we are one and the *same total physical universe*.

The difference in viewpoint between a Cosmic Unity Theory and the cosmological Zeitgeist regarding worlds outside our own is thus about as far apart as one can imagine, even though we both start with a basic proposition that recognizes the applicability of the same physical laws within our universe. Max Tegmark, of MIT, illustrates these poles-apart views when he states, "I fully expect the true nature of reality to be weird and counterintuitive, which is why I believe these crazy things."[30]

On the contrary, though, I believe that the true nature of reality in the gigantic parent universe that physically surrounds our universe will actually prove to be more intuitive than most cosmologists evidently imagine. I, for one, can intuit an extremely interesting parent universe possessed of the same general laws of physics as ours, recognizable in most ways and undoubtedly very surprising and astonishing in others.

As mentioned before, I don't see that cosmologists truly apprehend the possibilities we're discussing here—that is, a concept of a true, compatible, surrounding parent universe and the answers it offers. Alexander Vilenkin says that, "Were it not for the multiverse picture and anthropic arguments, we'd have no tools for thinking about the structure of the universe beyond our horizon."[31] Well, I'm offering here and now the perfect tool for thinking about the structure of the universe beyond our horizon. It's definitely not the multiverse envisioned by the professionals—one that includes, among other things, a cosmos with the incomprehensibly differing worlds possessing wildly differing physical laws, buried in infinities and perpetuities. Rather it embodies the same general relationship that exists in our universe between stars that collide, stars that go supernova, etc., and the remnants of any version of the foregoing.

For us, that tool sought by Mr. Vilenkin for expanding our thinking about the structure of the universe beyond our horizon is called the Cosmic Unity Theory.

As noted before, and as we shall particularize later, multiverse theory doesn't give us the answers we need over a rather broad range of individual points. The default assumption of Mr. Davies mentioned above was that, in the absence of a grand unifying theory, the universe is simply made up of a haphazard patchwork of universes with differing domains, properties and laws of nature. But CUT wipes away such default assumption by offering a *unified* structural theory, as opposed to such patchwork of various universes. To repeat: this other-universe-equals-probable/possible-different-laws-of-physics equation should not apply to a Cosmic Unity Theory because we're the *one, same total* universe.

As noted, Cosmic Unity not only represents a *structural* solution for our universe, but a *birth* solution that is equally simple, consistent, logical and fitting. There is no necessity for us to search for the weirdest possible concepts in order to explain how and why our universe came into being, and how we may be related structurally to a larger cosmos.

In comparing the simplicity of Cosmic Unity theory to the complexity and unsupported speculation of BBT and multiverse theories (as we shall soon delineate) one might even employ William of Ockham's admonition (borrowed from Aristotle) to the effect that in science "entities must not be multiplied beyond what is necessary."[32] For example, assume that you walk into your house one day and find that a half glass of milk you had left on the table was knocked over, breaking the glass. The milk is gone. You have a cat, and notice that your cat is licking its whiskers, seeming to have a very satisfied look on its face. On the other hand, though, the thought occurs that a burglar could have climbed in through a window in your house or picked your door lock, accidentally knocked over the glass of milk while he was searching your home, then mopped up the milk but left the glass in place. Nothing appears to have been stolen from your home and there are no signs of forced entry.

Do you think the cat or a bungling burglar committed the dastardly crime? If we follow Ockham's Razor we'll go with the CAT, not BBT (Bungling Burglar Theory). So also, if we follow Willie O's dictum regarding the origin of our universe we'll go with CUT, not BBT. (!)

When compared to the simplicity of CUT, all of Ockham's objectionable "multiplied entities" embraced by BBT—that are built with conjecture upon conjecture upon conjecture—would melt like a snowman before the warm rays of summer's sun. An objective examination of some of the strange BBT and multiverse concepts (to be discussed in Chapter 4) make Medieval discussions about how many angels could dance on the head of a pin seem quite pragmatic.[33]

It should be noted that some of the most popular multiverse, megaverse, pocket/endless/perpetual universe theories resort to concepts of infinity that essentially make anything possible, however improbable. Their multiverse belief in an infinite number of physical laws or an infinite number of universes carries with it notions that, for example, there is another *you* out there. In fact, there is an infinite number of *yous* out there, as well as an infinite number of *yous* with only slight differences (a birthmark on one arm; a *you* weighing 180 pounds vs. 210; a *you* married to Jane vs. Judith, *ad infinitum*). Such infinitely-premised multiverse theories directly raise the prospect that we could/would probably be living in a fake universe itself—a computer-generated universe, etc. After all, we're talking about an *infinite* number of universes, laws of nature and possibilities, right? Infinity, almost by definition, has to result in such conclusions. It's basically just another way of saying that anything that's possible, *is*, and it *is* in infinite numbers and variations.

The concept of "eternal inflation" in a type of bubble-bath universe of spontaneously-generating universes is another of the suggestions bandied about in this cosmos of infinities. Branes, dark energy, and concepts of beginninglessness and endlessness are integral parts of such theories.

Well, let's forget about such open-ended infinities and perpetuities for a moment.

Without such considerations scientists believe that the chances of another universe with compatible laws, as the quotations above demonstrate, are about nil. I am not necessarily disagreeing with hypotheses that other *unrelated* universes, if they exist, might well be subject to other laws of physics. Such may be true, as almost anything is possible. Also, some of the things that we believe to be fundamental laws of physics in our own cosmic family may be more accidental than fundamental, and could just possibly change from parent universe to offspring. But let's not always search for the strangest theories where such searches may not be necessary. It's fun to let the mind wander to all sorts of strange and wonderful possibilities about other universes (I definitely do some speculation later in this book); but right now we're talking about our universe and the parent universe that surrounds and embraces us. There is no necessity to have to conjure up pictures of the most extremely odd, shocking, unfathomable other universes and/or other dimensions when the answer to the origin of our universe could be so simple and apparent as CUT. I think the possibilities for surprise, shock, and speculation offered by our own real, extended, total universe under CUT will be more than enough to keep us ruminating for a long, long time.

4. **The cause of our universe's birth was probably an explosion of a star in our parent universe.** The first three propositions relating to CUT are the most important. They place us in the right universe, if you will, regarding the Big Question. Once we've accepted the fact that we're all a part of a larger, related cosmos, and the fact that our universe came into being as a result of some event in such next larger surround-

ing universe, the door is open to a multitude of follow-up questions that should occupy us for a very long while. I'll offer some of my thoughts as we move on from here, but I realize that there may be many people who could be much more capable of proceeding to the next level than I. It would be especially helpful, for example, to have at one's disposal software with which to make a variety of computer simulations involving our parent universe under differing assumptions.

With our first three basic propositions in place, here's one way to proceed:

We can first ask the structural question: How does our universe physically fit into our larger cosmos?

We have touched on this "location" question before. Under CUT we perceive the following:

Unlike BBT, which would have to declare that the only "location" for our universe is here, CUT is obviously not so confined. Our universe, under CUT, has particular reference to other objects in the surrounding medium. In some respects such next level of the surrounding universe can be spoken of and conceived of as a separate entity, and in other respects such an approach is quite misleading. Cosmic Unity holds that our universe is physically surrounded by our parent universe of which we comprise a part. Our universe is not a separate entity in that respect. Yet the explosion that brought our universe into existence clearly created a microcosm within the larger, total universe in a manner somewhat analogous to that which occurs within the blast field of an explosion here on earth. An exploding bomb here on earth can be said to create its own microcosm within the ambit of its blast field for some period of time and for some radial distance. So also the blast field and remnants of our exploded progenitor star created some temporary divergences in the immediate vicinity in our surrounding universe. Such divergences will dissipate more and more as time passes and our universe's expansion ever more results in assimilation of our SUSE's remnants into the medium of our surrounding parent universe. (We noted the splash-in-the-pond analogy before.)

It's directly analogous, if not mechanically the very same, as the supernova blasts or star collisions in our universe. For example, in our universe things are very different between the supernova's newly-created "microcosm" and the surrounding intergalactic medium for some time and some distance after the initial explosion. But in time things get back to normal, and the usual order of things that might be said to have been temporarily overcome in the immediate vicinity of the blast ultimately reasserts itself. The same thing could be said to be happening right now in our surrounding universe after the SUSE that brought our universe into being.

Under Cosmic Unity, the farthest point of our universe will not be a barrier, wall of a bubble, foam, brane, or even nonexistence. Our universe simply melds into the surrounding pre-existing cosmos.

One should not, by the way, believe that what we consider to be our universe constitutes the entirety of the matter that made up the huge progenitor star that exploded or crashed in our parent universe. If nothing more, I would at least envision a probable remaining core, in whatever form, existing outside the relatively small ambit of our universe. Cosmic evolution after a supernova or other star cataclysm in our universe is a common sight, and one would expect the same principle to apply in general throughout the cosmos. Some remnants in time evolve into large cosmic bodies, some into smaller ones, and some into scattered degenerate matter floating in space without consolidation into any large mass at all. In fact, as a consequence of supernova explosions in our universe some ejecta is occasionally blasted entirely out of the very galaxy of which it formerly constituted a part.

At the vastly differing scales in which our universe and the total surrounding parent universe are operating, it's altogether appropriate that our universe has evolved the way we see it to be under a Cosmic Unity concept. Even though our universe is located in a surrounding total cosmos in which giants also evolve and swirl it is perfectly natural that the portion of the total universe made up of our visible universe contains no monster celestial objects yet. CUT conceives that outside the parameters of our portion of the total universe there probably exist immense structures that are beyond any conception we have mustered, even though they are lesser in number than objects more within our size and usual scale of perception. I believe that the variety in sizes of structures in the surrounding medium of our parent universe will prove to be truly amazing.

Although we do see a great diversity of celestial bodies in our universe—and many of these we don't understand yet—it's my thinking that our universe is relatively non-diverse in comparison with the total cosmos that exists beyond our borders. I say this fully believing that we and our parent surrounding universe are probably possessed of the same basic laws of nature. But everything within our universe is especially closely related to us. After all, everything we see probably came from the very same progenitor star in the same explosive event in the same parent universe.

In the smaller niche of the total universe we know as our universe we can detect things the size of neutrinos all the way up to galactic clusters, walls and voids. Our surrounding universe encompasses all of that plus the enormously larger structures appropriate to the immensity of such larger total cosmos.

Once we accept that we came from a surrounding parent universe I think the way is open for us to consider a variety of occurrences, both natural and possibly non-natural, that could provide the basics for our universe coming into being, in addition to the SUSE we have discussed (and I encourage examination and analysis of the possibilities). But if for now we stick to our "most reasonable" and "simplest" explanations I believe our conclusion should be that our universe originated as a natural result of an explosive star event that occurred in our large, surrounding, universe. For purposes of our discussion I am including within the category of star explosion that brought our universe into being various types of parent universe star cataclysms. Although I feel that some genre of parent universe supernova definitely leads the list of possibilities, when I use the acronym SUSE my intention is to include a broad range of colossal explosions that could yield remnants of one kind or another.

I do want to make clear the fact that Cosmic Unity is not a version of a general idea that our universe might simply be a single atom in a larger universe, which is an atom in a larger universe, etc., with the atoms in our universe in reality being universes for smaller realms, etc. The mechanics, expansion, and evolution we see in our universe simply don't comport with such a concept, and such a concept is not what Cosmic Unity stands for at all. For example, under CUT our universe was created from an explosion. The SUSE that resulted in an expanding and ever-evolving universe is something that does not describe the life of an atom. Further, such an atom-of-an-atom-of-an-atom theory would undoubtedly fail because of scale confusion. Something that holds true at one scale will often not hold true if tried at a larger scale. There are also other and more technical reasons that would seem to rule out the atom-within-an-atom concept. One such example: the second law of thermodynamics, which deals with irreversible processes (like, say, a perpetual motion machine), would also appear to exclude the operation of such an infinite verticality of universes.[34]

CUT, though, is faced with no such problems.

In arriving at CUT we take notice of the very real, natural entities that are physically observable every day in our universe in the form of stars. We don't have to guess about what the stars in our universe are mostly made out of; we don't have to guess that star collisions and supernovae result in remnants; and we don't have to guess what develops from such remnants in our universe. We are quite certain about most of these matters. New stars, planets, and a wide variety of celestial bodies that inhabit the space around us in our universe eventually develop as the result of star explosions. Under CUT we don't have to resort to other theories' mysterious energies or matter or concepts that strain credulity and imagination. We *know* what happens when stars explode or collide.

After taking due note of the above, I ask the reader: Are you open to answering these two questions affirmatively: (1) Does the concept of a large, unified cosmos seem at all reasonable to me, and if so, (2) am I willing acknowledge the possibility that some type of aggregation of matter and energy (similar to what we observe in stars in our universe) might exist in a larger cosmos?

If you can answer affirmatively to the two foregoing questions we have the very foundation for our universe within a total larger surrounding universe under CUT.

•••••

What would be the characteristics of the star(s) involved in the SUSE that brought our universe into existence? The easy answer, of course, is that they were the characteristics that were necessary to make us what we are.

In Chapters 4 and 5 following I will offer the rationale why a SUSE offers the more reasonable basis for our universe's origin as compared to BBT. In Chapter 4 we'll see why the major inconsistencies and obstacles of BBT are either wholly solved or very substantially reduced with CUT. In Chapter 5 we show that the arguments proffered *for* BBT apply with the same or even greater force to a theory embracing Cosmic Unity everywhere. The fact of the matter is that our universe *did* result from a bang of some nature, but it was not the weird singularity that did not exist in time or space, and that supposedly was the first thing ever to exist anywhere, and which violated the laws of nature by mysteriously appearing from nonexistence for no known reason. On the contrary, the celestial bodies that comprise our universe came into being in much the same manner as our local stars and own solar system's inhabitants came into being. As remnants of a larger exploded star.

Some people speculate that black holes might somehow spawn other universes. Lee Smolin has proposed that black holes could be a source of unseen new universes, each of which begins with some type of a big bang, and each of which has physical laws differing from the parent universe. One of the main reasons for suggesting a black hole-universe-creation theory evidently arose from the fact that Smolin observed that black holes might be able to establish an alternative to the type density that only the Big Bang could otherwise create for the cataclysm that began our universe. But CUT suggests three things: (1) Why search for a universe-originating entity (ie., a black hole) that would have to change its stripes to create new stars, planets, etc. when we have the tried and true template for universe creation (ie., supernovae) staring us in the face on a daily basis? (2) Why not conceive of a "child" universe with compatible physical laws, instead of an offspring with differing laws from the parent? and (3) Why not go cosmic in search of a big enough entity to bring our universe into being? Under CUT our SUSE, which involved the implosion/explosion of a star in our parent universe with the mass of more than ten sextillion times the mass of an average star system in our universe, is the natural answer.

Our SUSE might, incidentally, have left behind a black hole, but it is the expanding/evolving remnants of the exploded star to which we owe our universe's development, not to any remaining core in the form of a black hole.

Stephen Hawking talks in terms of possible "baby universes" made up of particles that can escape from black holes in different forms in which they entered.[35] (For this reason he suggests that astronauts use another mode of travel than black holes. He points out that space travelers would enter a black hole in real time and exist in sticky parts in "imaginary" time. Good point.) Such speculation is interesting, but black holes don't seem to equate into things we normally associate with expansion, evolution, development, or even life itself. We usually equate black holes with just the opposite. However, the remnants of supernovae, for example and for good reason, are associated with those attributes which are absolutely proven to be helpful and even necessary for the creation of new worlds and new life.

It could be, then, that in the past we have been looking for the right answers to the Big Question in all the wrong places. No theory relating to the origin of our universe other than CUT (1) solves the many serious problems facing BBT, (2) at the same time incorporates all the BBT proofs offered for a "bang" into a cohesive all-embracing theory of its own, (3) embraces most of the essential premises of the Standard Model of BBT (without adopting many of the incorrect conclusions people have drawn from them), (4) relies on a simple, reasonable explanation based on a larger-magnitude happening of something that we see as a common occurrence in today's universe, and as such (5) is based on something (the explosion of large stars) that we are absolutely sure has a proven track record of creating real (not hypothetical) stars, planets, moons, asteroids, comets, and the host of other celestial objects we see before our very eyes every day.

4
Problems With Big Bang No Problem For Cosmic Unity

There are many evidentiary arguments for CUT's position that our universe originated as remnants of some type of explosion involving a progenitor star or stars in our large, and surrounding, universe. As noted in the last chapter, much of such evidence for Cosmic Unity fits into two general categories.

The first category is based on the very arguments that are advanced to support BBT. After all is said and done, certain foundational premises of CUT—just as with BBT—are that our universe started with a bang, the results of which are expanding and evolving. Many scientists and non-scientists feel that, despite BBT's problems, no other theory for the origin of the universe even approaches the strength—and resultant scientific acceptance—of BBT. With that in mind I feel that if one compares CUT to top-ranked BBT on a point-by-point basis, people will come to understand how CUT clearly prevails as the favored answer to our origin-of-universe question, co-opting BBT's own proofs in doing so.

The second general category of proof for CUT involves directly facing up to the problems inherent in BBT, noting that those problems of BBT disappear if the "bang" that created our universe came from a parent universe star explosion instead of the Big Bang.

Therefore (1) CUT not only possesses all the proofs[1] usually presented in favor of BBT (some of which are more convincing for CUT than BBT), but (2) as a big bang theory the SUSE of CUT also solves the inconsistencies and contradictions inherent in the impossible Big Bang of the impossible beginning singularity.

Let's venture into the second-mentioned course of proof first, taking cognizance of the fact that the problems of classical BBT disappear if our universe originated with a SUSE under CUT. In looking at some of the problems plaguing BBT we should note that the list of BBT critics is long and distinguished, even though such list comprises only a very small percentage of the total scientific community. In 2004 dozens of scientists, engineers and researchers published an "open letter" to the scientific community observing that BBT has to rely on so many "fudge factors" that without such "growing number of hypothetical entities" BBT would be fatally wounded.[2]

•••••

The following, then, are some of the biggest problems with BBT—all of which go away under CUT.

A. The Big Bang theory has to hold that we—and everything else—came from nothing. We have already quoted some of the surfeit of statements from BBT supporters to the effect that everything came from the Big Bang and nothing preceded it. There have been all sorts of attempts to hypothecate some type of very, very, very, tiny originating Big Bang embryo, but there is always the ultimate unanswered question for BBT: Well, where did *that* originate? Then where did *that* originate? Etcetera. Even if it was only some type of strange seed energy that supposedly marshaled itself, or a quantum fluctuation hiccup that started it all, the questions remain: Where was that energy *located* before the Big Bang; what was *it* made of, and what started *it,* and why did it suddenly launch itself into existence? The same questions hold true for various "vacuum" and other theories. By claiming that the Big Bang created time and all matter BBT necessarily holds that nothing existed before the Big Bang and that the "nothing" we're talking about wasn't located any place.

But then our whole universe came from it, right?

The SUSE does away with those clear violations of logic and the laws of conservation of energy and matter. If our universe originated from a cosmic unity SUSE we came from a definable, real object or objects that possessed all the elements necessary to have created—as we ourselves observe as an ongoing process—the many quintillions of new structures in our universe. As mentioned before, large exploded stars have a proven track record in creating worlds of new structures, if you will. As previously noted, in our visible universe, for example, supernovae alone occur dozens of times a second; and a supernova of the type that brought our solar system into being occurs as often as every 30 years in just our galaxy alone.[3] Our solar system could be Exhibit A illustrating what the remnants of a (possibly second or third generation?) star that goes supernova in our universe can do in about 5 billion years or so. So, the star that exploded in our parent universe, as a creature of such surrounding universe, simply did the same thing that in time caused our moon, earth, sun and galaxy, etc. to come into being, but on a larger magnitude (with all appropriate cosmic evolutional and progenitor star "generational" adjustments accordingly made). CUT maintains that our universe is made up of those remnants of the SUSE, some of which are visible to us, and some of which are beyond our visible horizon.

And please, for those who are considering CUT for the first time: Don't look to the quantum world to supply a logical alternative for the origin of our universe. Blithely saying that things temporarily jump in and out of existence in the quantum world, and that our entire universe could have originated somehow in the same way, is absurd and, in the vernacular, a cop-out. Things don't jump in and out of existence in the quantum world in the sense that matter is created from nothing, and certainly not worlds of matter, and certainly not permanently. Maybe we can't always *measure* things, or are uncertain about the precise location of particles (or strings), at any given moment in the submicroscopic quantum world, but to use inapt speculations about quantum physics as an explanation for the creation of all matter, energy and time in our entire universe requires a total discard of reason and experience.

Cosmic Unity, on the other hand, solves the implacable everything-from-nothing conundrum facing BBT.

In the end, humor and irony manage to find a way to pop up in almost every aspect of life. Cosmology is no exception. The small percentage of scientists who oppose BBT are adamant, and at times bitter, in their denunciations BBT. Their arguments against BBT come at BBT from almost every conceivable angle, from the more obvious to the most abstruse and arcane. But they don't mention the 800-pound gorilla sitting in the cosmology room—the fact that BBT essentially says that it self-created itself from nothingness, from total nonexistence. Any objectively critical view of BBT would have to be that it is most vulnerable in staking out a position that everything that is or ever will be came from nothing—At least that position would have to be right up there in the number one or two slot. So, why don't the anti-BBT scientists attack such a vulnerability with the biggest guns possible? After all, the everything-from-nothing concept violates the cardinal rules of physics. Well, it turns out that most of the anti-BBT scientists *can't* go after BBT where it's most vulnerable. Why? Most of the scientists in the anti-BBT clan still believe in SST, and the Steady State Theory is even a worse violator, if that's possible, than BBT on the issue of creation from nothing. Thus the laws of conservation of matter and energy can't be invoked against BBT by those who would probably like to most invoke them.

BBT party members may well despise me for the things I say about their theory, but it unfortunately seems that the SST guys and gals won't have much better to say about me. It's the burden and opportunity of CUT to alienate both the majority and minority parties in this game of cosmological politics on the big questions relating to the origin, structure and future of our universe. Maybe it's time to vote into office an independent cosmological candidate by the name of CUT.

B. Time didn't exist before the Big Bang. Big Bang proponents must carry the uncomfortable and inconsistent baggage that goes with a theory that says that nothing at all, including time itself, existed before the Big Bang inflation began. Some Big Bangers find it hard to accept such a position, but it's part and parcel of the Standard Model BBT. Since we now talk in terms of space and time being connected in a space-time continuum, we really can't separate the two. Thus if Big Bangers continue to argue that no thing (which includes space) existed before the Big Bang they of course are stuck with contending that time didn't exist before the Big Bang either.

Time is said to be our fourth dimension. Among the many time-related questions I have for BBT: if time began in a closed universe with the Big Crunch singularity that expanded at the Big Bang, what happened to the *time* in the previous Big Bang/Big Crunch cycle(s)? Did time actually not exist for the prior 100+/- billion years? Even though a universe existed in the prior cycle? Are we left with saying that all past time didn't exist? And that time is starting over? And the things that transpired during the prior time didn't happen after all? Or is time really just continuing and NOT beginning, as claimed under BBT?

Under BBT do we have two different scenarios for time, depending on if we have an open or closed universe? For example, does time die under a closed universe, but live on in an open universe—because if space in a closed cyclical universe periodically disappears into zero, whether for a trillionth of a second or for millennia (as some have suggested), does time also disappear? And once time disappears does it thereafter (and I should not even be able to say "after") start over again? Is the status of time in the end, then, going to depend on how much matter there is in the universe under BBT (because if there isn't enough matter, the universe is thought to continue to expand forever, and if there is more matter it will in time compress back into a Big Crunch)? That seems more than a bit odd to me.

The foregoing questions don't even begin to touch on the myriad time issues under BBT. But if our universe originated with a SUSE (or some other occurrence in our parent universe) these time questions and many, many other related questions are answered; and of course under all such alternate scenarios the answer would be that time *did* exist before the Big Bang. Of course it did. It existed in our parent universe of which we are a part.

Which "time theory" seems more reasonable to you?

C. There is nothing on the outside of our universe. Various issues are involved in the position of BBT to the effect that when our universe expands it does not expand into anything. In fact, even use of words implying that we can move "outside" our universe or "into" something else is inappropriate under BBT. This entire broad subject of discussion involves questions relating to space, time, aspects of expansion, boundary concepts, and considerations relating to existence and nonexistence before and after a big bang. I have the strongest possible disagreement with BBT in these areas, which are rooted, again, in BBT's premise that everything began with the singularity's Big Bang, and nothing existed before it.

Our universe is proven to be expanding, but Big Bangers have never really been able to satisfactorily tell us what it is expanding *into*. Big Bangers argue that there is nothing outside our universe; that our universe has no boundaries or edges; that there *can* be nothing outside our universe; and that it is possible for our universe to expand without moving "into" anything, including nothingness. However some anti-Big Bangers argue that our universe does have edges or boundaries, and that expansion is impossible unless there is something finite to move into. (They don't know what it is, but they say that if our universe is expanding it must expand into finite existence.)

Under CUT Big Bangers are wrong on the question of whether there is anything outside our universe. There is something outside our universe, and it is the space of our surrounding parent universe. It should be pointed out that it is the BBT position, not the CUT position, that flies in the face of all reality and experience regarding the question of what our expanding universe is expanding into. It is BBT, not CUT, that seeks to carve out an exception to the rule that everything that expands or moves, moves into *something*. CUT says that our universe expands or moves into existing space, just like everything else we have ever seen. *Everything* else. Yet BBT says that, not only does our universe *not* move into existing space, but that it is up to others, and that would include CUT, to prove—by measurement—that BBT's stated exception to what we have always seen and experienced is wrong.

Of course the requirement of proof should be the other way around. It should be incumbent on BBT to prove, by physical "measurement," that its proposed exception to all other reality as we know it applies in the one instance of BBT. And such proposed exception just might apply.

It *might*. But until proof is proffered, CUT should seem to be much more reasonable than BBT in its fundamental position about what our universe is actually moving into.

I believe that Big Bangers are also wrong that our universe has no boundaries or edges. Our universe has boundaries in the sense that there is a physical limit to what constitutes our universe. Under CUT such boundaries exist in the same way supernovae in our universe have boundaries. As the remnants of supernovae spread out farther and farther such boundaries are in time not always easily detectable, but this is a matter of measurement as opposed to a matter of nonbeing.

In my opinion Big Bangers are further wrong when they say that there *can't* be anything outside our universe. It's true that there can't be anything outside our universe under BBT. As we know, anything outside our universe sounds the death knell for BBT because then something pre-existed the singularity, and as a result BBT didn't start everything. But of course the entire concept of CUT is that there is and was a pre-existing total universe that surrounded our SUSE.

But then comes the interesting point. BBT holds that it is possible for our universe to expand without moving "into" anything. Some anti-Big Bangers, though, say that it is not possible to expand unless one moves *into something finite*.

In the original draft of this book I devoted considerable effort and quite a few pages presenting the anti-BBT argument and BBT's pro-expansion argument on this point. What I came to realize, though, was that I was spending much too much time tilting at cosmological windmills. If one believes in CUT one doesn't have to face up to the question of whether it would be *possible* under a BBT scenario to, in effect, expand into and set up housekeeping in nonexistence. Such a question is entirely moot for CUT. Since that is the case, I finally decided not to get entangled in an issue so distracting to CUT. My short book thus became about 15 pages shorter. I'll just briefly touch on the subject.

Big Bangers present several arguments why they feel it would not be possible to move out of our universe, such arguments ranging from a premise that the shape of space won't let us move out of it to one based on our inability to see or measure anything outside our universe.

As mentioned a couple paragraphs above, in my first drafts of this book I found myself presenting extensive counter arguments to the several reasons asserted by Big Bangers for the proposition that when our universe expanded it did not have to expand into anything. I thought the BBT *reasons given* for the perceived ability of the BBT universe to expand were wrong. I began dealing with such reasoning almost ad nauseam. For example, I didn't think that the shape of space should be given as a reason to counter an argument that the universe must have something to expand into. No matter what the shape of space (is it flat, spherical or saddle-shaped?) an expanding universe increases its total volume, and arguing that our universe doesn't expand "into" anything based on the shape of space avoids the point at issue, in my opinion. Incidentally, Brian Greene's *Fabric of the Cosmos* (pages 238-250) gives a very good discussion of what are the modern views on the shape of space, and Chapters 10 through 15 of Levin's *How The Universe Got Its Spots,* gives a detailed theory of space-shape that is more than enough for most of us.

One solution-by-curvature argument employed by Big Bangers is that Einstein's concept of general relativity suggests four-dimensional (versus three-dimensional) space. They say that such curved space resolves BBT boundary problems. Here's how best-selling author Timothy Ferris describes it. General relativity, says he:

> ....resolves the paradox of whether there is an edge to the universe, by raising the prospect that cosmic space has an overall curvature.... Such a universe is finite, since it contains a finite amount of space, but unbounded: One can see forever, or travel indefinitely far, in any direction, without ever coming to an edge.[4]

The above words were written in the late 1990s. Now, though, such resolution-by-curvature is apparently already out of favor under the latest analyses. In fact, some scientists insist that the Wilkinson Microwave Anisotropy Probe (WMAP) in 2001 conclusively shows that the shape of the universe is not curved, but flat. Curvature thus may not be an answer to BBT's quandary about what our universe is expanding into.

Also, there is the argument presented by professor Ned Wright and other BBT advocates that since we can't see or measure anything outside our universe there's nothing "profitable" to be gained by even posing a question about what our universe is expanding into. That position amounts to nothing more than an avoidance of the issue. Originally I went into great detail attacking that position on several levels—in fact pointing out how BBT itself flunks such a seeing-or-measuring litmus test as an explanation for the origin of our universe, while CUT passes it with flying colors. But, as mentioned, I abandoned a several-page commentary on that BBT-based argument because of mootness.

So, this is all I'll say about whether our universe must move *into anything* when it expands. It's my thought that in theory it might well be possible for a hypothetical universe other than ours to expand without moving into anything finite. I say "in theory" because the situation definitely doesn't arise for our universe. I'm not going to spend my time here arguing against the reasons advanced for a proposition that might be right (1) for the wrong reasons (2) for a universe that's moot under CUT. Our descendents will just have to face up to that question much later as it may pertain to our parent universe.

The argument that a hypothetical universe might be able to expand without moving into something finite could be based on the concept that space comes into existence when and as there is a presence of matter or energy. Space thus becomes defined only by what's in it. The distance, or "space," between the various particles of matter or packets of energy is not and does not need to be any *thing* finite. Period. Nonexistence, or nothingness (which is the hardest concept for this writer to envision), doesn't enter into the equation. Since nothing is nothing it is also not a barrier to outward movement of things in an expanding universe. Parenthetically, I'm not sure I agree with J.B.S. Haldane's expressed suspicion back in the 1930s to the effect that "the universe is not only queerer than we suppose, but queerer than we can suppose," and Sir Arthur Eddington's similar expression that "Not only is the universe stranger than we imagine, it is stranger than we can imagine." I do suspect, though, that *nonexistence*, or nothingness, is queerer and stranger than one can suppose and imagine.

Phrases such "empty space" or "finite space" often only add to the confusion.

In any discussion about whether it is possible to expand without moving into anything a beginning distinction should always be made. Are we talking about our universe, or some hypothetical universe? Big Bangers could be right in a belief that such a thing might hypothetically be possible, but they're wrong in trying to apply such an idea to our universe—which is exactly where they intend that such a belief applies. Our universe has always been expanding into something. So, this is not an issue for us in a CUT world, as our SUSE and its expanding remnants have always existed in the milieu of the existing total universe.

Even in noting that it might well be possible to expand from a theoretical universe without moving into "something," I shall nonetheless acknowledge that there is also at least the possibility that if there is such a thing as nonexistence it *could* be absolute and implacable—and in such eventuality the famous balloon we often talk about could never expand past the implacable barrier of nonexistence. Atoms, quarks, whatever, couldn't move into nonexistence or they themselves would not exist. Nonexistence, in such a prospect, would not be occupyable.

But our universe has never encountered nonexistence (for one thing, there's nothing to bump up against) nor ever had any way of encountering it.

So let's move on.

•••••

Comments about expanding "out of" our universe or across "boundaries" of our universe, etc. violate the BBT credos in this area. The alternative to committing such violations could be to join in intellectual exercises that hypothecate that our universe is the lone known exception to requirements that things have edges or boundaries of some kind. But in refusing to join in such hypothecations it's not that people who disagree with BBT haven't heard the BBT positions on the above issues; it's just that such definitions and basic positions are directly challenged.

It's also not that people couldn't possibly dream up even more exotic explanations than BBT for how it might be theoretically possible for our universe to come into existence, expand, create itself, etc.; it's just that some people choose not to follow the route of highest speculation, complication and uniqueness for our universe.

Maybe it's time to say that enough is enough. Perhaps the time has come to challenge the BBT credos and accompanying language in this area. Regarding the issue of this section people could therefore bring themselves to say: Our universe is finite, just like everything else. It has boundaries, just like everything else. When something passes those boundaries it moves into finite space, just like everything else. That finite space happens to be the realm of our surrounding parent medium.

Could some type of hypothetical universe *possibly* expand without expanding into something? Maybe, but we're talking about our universe now, and that question doesn't apply to us.

D. More time-related problems. We have mentioned a few time-related problems with BBT. Some critics of BBT point out additional difficulties with BBT: For example, some astronomers have observed (1) apparently very old galaxies—seemingly too old to fit into a BBT scenario—and (2) huge structures of multiple galaxies in the form of gigantic strings, walls and voids that may have existed for many billions of years before they should have had time to form under the zero-size-to-larger expansion envisioned by BBT.[5]

At present the issue of how such facts of structural cosmic life came to be within the time allotted to it under BBT is without adequate explanation for many people. Expansion velocities faster than the speed of light would have to have been required to achieve such structures and voids, according to most scientists. Whether "inflation theory" and a beginning repulsive negative force that conveniently turned on and off again within a billionth of a billionth of a second (or so)[6] will again ride to the rescue of BBT on some of these points is unknown. Many people feel that arguments based upon fine-tuned on-off expansion speeds of the universe according to an "inflationary" addendum to BBT are made in attempt to save the Big Bang as a theory rather than to approach the conundrum objectively.

But look at the much more reasonable alternative.

What if our universe started and spread out from an object that had structure itself? What if that structure might have been incredibly huge in comparative terms to BBT's pinpoint—possibly measuring trillions of miles in diameter at time of explosion?

Such a beginning structure could have been our SUSE, and such a structure could provide some answer to the astronomers' question about how the large-scale cosmic structure we now observe in our universe seems to have had a head start in size development that is inexplicable under the timeline of BBT without intervention of the physical miracle of faster-than-light-speed inflation that had to turn on and off within less than a billionth of a billionth of a second to work.

There is also the so-called "horizon problem" that raises concerns about time and assumed beginning size of the universe under BBT. Most scientists believe that the "smoothness" (the uniformity of microwave background radiation) in space can only be explained if the photons after the Big Bang had been thermalized (very thoroughly mixed together) by early particle collisions. But our horizon size (the distance photons can have traveled while our universe has been expanding) is said to be too small for our universe to have developed in the time necessary to achieve our universe's smoothness—if in fact our universe started out as a tiny pinpoint of nothingness.[7]

At first, the horizon problem was acknowledged to be a formidable barrier for BBT to overcome. But BBT adherents have, as observed a few paragraphs above, come up with the strange accommodation called "inflation." This phenomenon, originally proposed by physicist Alan Guth, has now been theorized into the new, improved version of inflation, often referred to, not surprisingly, as "new inflation." New inflation is thus depended on to solve the horizon problem (as well as the developed structure problem referred to earlier.) Again, this horizon problem is caused by the apparent uniformity of temperatures at distances too far apart in the universe to have communicated with each other at any time. Coincidences in the form of such uniform temperatures just don't occur in nature. BBT believers contend that the "new inflation" of our new universe did the following: (a) started the expansion of our universe at a rate within the laws of nature; (b) suddenly speeded up the expansion to an unheard-of rate, and (c) then suddenly slowed the inflation down (d) at precisely the needed rate in order to conveniently answer the horizon problem. Here's the way Brian Greene explains how this physical marvel worked under the newest BBT version of inflation:

> In a nutshell, space expands slowly enough in the very beginning for a uniform temperature to be broadly established and then, through an intense burst of ever more rapid expansion, the universe makes up for the sluggish start and widely disperses nearby regions.
>
> That's how inflationary cosmology explains the otherwise mysterious uniformity of the microwave background radiation suffusing space.[8]

To me, the unerringly precise sluggishness, then speeding up, then slowing down of a *physical act* called "inflation" (remember, we're talking billionths of a second here) that just happens to supply the exact time solution to the horizon problem seems more mysterious than the "mysterious uniformity" it hopes to explain. Some critics of inflation argue that if inflation occurred as suggested by Big Bangers it would have required a density some 20 times larger than implied by Big Bang nucleosynthesis (the process by which the elements—in this case the lightest elements—were formed by atomic fusion immediately after the Big Bang). I'm not knowledgeable enough on the point to make an educated guess, but many scientists hold such a conviction.[9]

From what I can see most BBT adherents admit that difficulties such as the "horizon problem" would in fact be just that—a huge problem—for them except for the saving theory of inflation they have come up with.[10] If I were a committed Big Banger, though, I do believe I'd try to find a solution to the horizon problem that didn't depend on what amounts to another physical miracle. Inflation theory will not go down without a fight, though. Some Big Bangers feel so strongly about it that they consider it to be the most important and lasting contribution to the field of cosmology in decades.

I feel we do run into the same type of circular reasoning as inflation is thought to relate to the horizon problem as we've seen advanced elsewhere by Big Bangers. For example, in discussing why the horizon dilemma is no longer considered to be a problem one staunch supporter of BBT definitively states that the horizon problem (and the "flatness" problem) "has been resolved by the concept of inflation." However, in the very same paragraph in which he declares that the horizon problem has been resolved by inflation, he then goes on to say:

> Inflation is a very rapid expansion of space by a factor of 10^{28} that allows the entire observable universe to have been in causal contact before inflation. *The main evidence for inflation is in the homogeneity of the CMB and in the flatness of the geometry of the universe, neither of which can be adequately explained without an inflationary epoch.*[11] [Emphasis supplied]

Let's follow the reasoning here. The homogeneity and flatness problems are claimed, in so many words, to have been "resolved by the concept of inflation." And what is the main evidence that supports the concept of inflation? Well, it's the homogeneity and flatness of the universe...."neither of which can be adequately explained without an inflationary epoch." So, problem 1 is "resolved" by problem 2; but problem 2 depends for its resolution on problem 1. In debate this error is called *circulus in demonstrando*.

Though other Big Bangers may not make the mistake of admitting so directly in the very same paragraph the dependence of proof for inflation on the very problems it purports to solve, and vice versa, it's implicitly there nonetheless.

The concept of inflation has become so imbedded in BBT that Andrei Lunde claims that Big Bang theory has now become part of inflation theory, not the other way around. He now draws a distinction between the BBT of old and the new inflationary theory of which the Big Bang is just a part. At his Stanford University Department of Physics website he states:

> Inflationary theory describes the very early stages of the evolution of the Universe, and its structure at extremely large distances from us.
> ....
> Initially, inflation was considered as an intermediate stage of the evolution of the hot universe, which was necessary to solve many cosmological problems. At the end of inflation the scalar field decayed, the universe became hot, and its subsequent evolution could be described by the standard big bang theory. Thus, inflation was a part of the big bang theory. Gradually, however, the big bang theory became a part of inflationary cosmology.[12]

So, the inflationary tail is now thought by some to be wagging the universal dog.

On that point I'll just finish with this: If the horizon problem truly does remain an unresolved dilemma for BBT *but for* the saving intervention of inflation, then the origin of our universe as the result of a SUSE could give the much more reasonable alternative solution. A SUSE (1) suggests an entirely different *mechanism* for our universe's smoothness (if our universe really *is* smooth, which some people deny), and (2) shows that the origin of the radiation was from a fantastically *larger physical area* than the almost-zero size of BBT.

Under Cosmic Unity the "thermalization" that is needed to solve the horizon problem occurred because the thing that exploded was a pre-existing, discrete, already-thoroughly-mixed object (a big progenitor star), *not* because of the operation of another one-time physical miracle.

Cosmic Unity doesn't need the physical miracle called "inflation" to explain thermalization.

E. Some singularity problems. Mathematicians refer to an event in which density and temperature are "infinitely high"—such as at the very beginning of our universe under BBT—as a "singularity." The position is taken that the laws of science do not apply to a singularity, so one really can't question any calculations or measurements used to identify a singularity (not that I personally would be capable of challenging the mathematical calculations and physical analyses that would be involved. But others would.)

However, I, for one, seldom ever use the word "infinity" or "infinite" to describe anything physical, so I immediately start out with a prejudice against singularities. I'm fully cognizant, though, that it's almost impossible to read a page relating to cosmology, or a paragraph relating to quantum mechanics, that does not contain some reference to "infinite" or "infinity." With that in mind I'm not about to make a stand against the existence of singularities based upon their continual reliance on the word "infinite" in some form.

Secondly, though, singularities have other problems. As soon as we are asked to suspend belief in things that apply in all other instances (here, to suspend logic and science) one has to be suspicious. So, I'm suspicious. I do have an open mind on the existence of some types of singularity, though, and if people believe that what happens in a black hole is a singularity, then that's what they believe and that's what it's called. I don't even have a problem if people want to refer to what happened with the SUSE that brought us into existence a singularity, but the definition is going to have to be stretched to fit what actually happened under CUT, not the other way around.

But so far as our *universe's* beginning as the oft-described singularity questions have been raised in addition to the fact that the singularity that allegedly brought everything into existence is said to come from nothing; that it never existed any *place*; that it created time and space; etc. Among other issues, some critics of BBT's singularity point to serious problems in respect to such matters as gravity, energy requirements and assumptions regarding velocity of expansion of matter immediately after the Big Bang (much above the speed of light, as noted) under most versions of BBT. Just how it would be possible for the universe to expand from a pinpoint of nothingness out to its present assumed size (even aside from the developed *structure* problems mentioned previously) in the required time has never been satisfactorily answered without a reliance on the previously-mentioned white knight of BBT called inflation

Concerning the time-related expansion dilemmas created by BBT's singularity, I as a nonscientist would again inject the same point made regarding BBT and our universe's highly developed *structure* problem: The indicated expansion of something (ie., a mega-monster star in our surrounding universe) many octillions of times larger than BBT's tiny starting pinpoint could take far less time to stretch to our universe's present size than if it expanded from its claimed zero size. Look at the headstart it would have had (though respective velocities of expansion could be a factor). The BBT party line essentially seems to be that (1) we should simply gloss over the physically-impossible facets of BBT (i.e., that require expansion speeds faster than the speed of light) and proceed to the next issue, or (2) we should in fact believe that inflation did occur faster than the speed of light for the fine-tuned, convenient period of time necessary to save BBT, but not be given any proven reason how or why.

There is of course this suggested reason offered by Big Bang believers for our universe being able to expand faster than the speed of light during an inflationary period: Einstein's injunction against moving faster than the speed of light only applies to objects moving through the space *within* our universe. The universe itself can effectively break the speed barrier because the newly-created space of our expanding universe is not the result of expansion *within* our universe itself.[13]

Just think of that explanation. Lo and Behold—evidently our universe is allowed to expand much, much faster than the speed of light because there is a distinction that is drawn between being either *within* our universe or effectively moving *outside* our universe (but just for this evident one-time limited purpose of saving BBT by way of inflation).

I disagree with such proffered explanation for inflation's expansion, but I admittedly may be wrong as to the possibility for such an event to occur in another universe. The very people who stoutly insist that there is no "inside" and "outside" of our universe—and that our universe has no boundaries—now find it very handy to ignore that specific dictum if necessary to save inflation as a theory. And inflation, as we know, is needed preserve the Big Bang model as a theory. Suddenly an end run—one that depends for its existence on being able to violate a cardinal rule of BBT and essentially reach *outside* the *boundaries* of our universe at an early time in our universe's existence—makes perfect sense to them if it saves BBT. Exceptions to exceptions to exceptions in order to keep BBT alive.

The above concept, however, may be right for a different universe. It just shouldn't be applied to our universe, in which all expansion takes place *within* the confines of the surrounding, pre-existing cosmos. Like our prior discussion relating to whether our universe is expanding *into* anything, the concept of extra-universal expansion at faster-than-light speed is moot for our universe. Even if the theory might be right for a different universe, we can put up a "Does Not Apply" sign for our universe.

CUT and BBT will just have to agree to disagree on whether our universe, whether inflationary or not, expanded into anything. BBT says we did not in the past and are not now expanding into anything; CUT says we did in the past and are now expanding into something. That "something" is preexisting space occupied by matter and energy located in the surrounding cosmos.

Further, the miraculous, perfectly fine-tuned rate changes in what is claimed to be the physical act called inflation are beyond the pale for this writer.

As previously noted, we don't know if the latest of the latest models of "new" inflation will continue to be considered sufficient to explain all the above-mentioned BBT problems raised by the singularity concept and its immediate aftermath. To this writer, however, an actual SUSE starting location, possessed of huge size in comparison with BBT's singularity, is much better adapted to accommodate the needed size, space, and time requirements inherent in the expansion that took place post-bang than the BBT's starting "non-location" of no size at all that must depend on so many exceptions to common sense, to the laws of physics, and to BBT's own usual tenets.

There are also singularity-related problems concerning the fact that certain proofs of BBT rely on the assumption that the universe contains only so much matter as we can see.[14] Such a postulates now seem to be false.

There also doesn't seem to be a reasonable explanation for the singularity's sudden decision, and ability, to begin expanding. Does anyone have any suggestions? Not really?

CUT, though, has the complete explanation.

Gravity is another problem confronting BBT's singularity and its expansion. We believe we know what happens when large-mass stars collapse. Gravity holds everything, even light, from escaping from the black hole thus created (though Stephen Hawking thinks that black holes may have backdoor leaks). How in the world (again, pardon the expression) could the mass of an entire universe escape from the stupendous gravity that would exist from the densest possible pinpoint conceived of in a singularity?....And do it every time a Big Crunch/Big Bang cycle occurs? Exotic "bounce theories," repulsive gravity theories and other strange concepts don't seem to satisfy many people, to say nothing of the fact that they rely almost entirely on speculation relating to concepts that only apply to BBT.

Regarding "bounce" theories, by the way, a problem obviously arises if the Big Bang is the very first or only "bang." What would the first and only singularity bounce off of? We know that the unimaginable density and resulting intense gravity of BBT's singularity would be an attractive force, unlike the negative force that creates the explosive expansion for a supernova.

One should highlight the difference between BBT and CUT in this important area. Recognizing the gravity-related dilemma facing them, BBT apologists have had to try to come up with some mechanism in an effort to save their theory, and the best they've come up with at this stage is the above-mentioned "bounce theory". Really, what this amounts to is an attempt to invent some unique process for the singularity (again) that already exists in the case of supernovae. Scientists are quite convinced, for example, that they understand the supernova process. Big stars in our universe exhaust their hydrogen fuel and implode, causing an intense rise in temperature at the core of the stars. It is said, for example, that the nuclear material of a collapsing star in our universe implodes to a density of almost two trillion pounds per cubic centimeter before it goes bang.[15] The result is an extraordinary explosion, with the outer layers of the stars being blown out into space in incredible bursts of energy.

Of course neither a beginning singularity nor a Big Crunch/Big Bang singularity is built for such a process, so BBT has had to invent one more adjunct (a "bounce") to its theory, an adjunct that applies nowhere else.

The logical question we should ask ourselves is this: Since in a supernova we already have a tried and true process for avoiding the otherwise-overpowering gravity that should come to bear in such a gigantic implosion, why try to mimic it with a made-up theory that is bound up in total speculation and applies only to the Big Bang?

Instead, shouldn't we simply go with a big supernova called SUSE?

•••••

All of BBT is not lost if CUT triumphs as the model for the origin of our universe. In building a model for the origin of our universe under Cosmic Unity a number of attributes of BBT can be incorporated.

For example: It's reasonable and provable that our universe emanated from a hot big bang. The Standard Model BBT is based on the occurrence of such an event. Let's fully accept a "bang" in the CUT model.

Second, under BBT expansion followed the hot bang. Let's accept that into the CUT model.

Third, evolution of our universe has been going on as it has expanded. Let's accept most of that.

But then the conclusions that BBT scientists have drawn from the foregoing post-bang concepts begin to get mired in establishment procedures and positions that are unsupportable. As just one example, in looking backward in time from today toward our beginning we simply should not try to trace our universe's lineage back to the singularity's impossible speck of nothingness. There is no reason for a trace-back to a singularity of zero size, to say nothing of the fact that such a procedure creates all sorts of other problems for BBT believers (and probably multi/megaverse believers that incorporate BBT singularity-type beginnings into their models). The basic procedural concept of a trace-back-to-zero is fatally misconceived.

There is also BBT's practical trace-back problem relating to maintenance of a correct balance of density to prevent either collapse of our universe or outward expansion into a virtually vacant universe during the past 14 or so billion years of an expanding universe. Some scientists claim that if we were to trace our universe today back to BBT's version of the Big Bang our universe would have had to have a beginning difference between actual density and critical density of no more than a trillionth of a trillionth of a trillionth of a percent to have avoided either a collapse or dilution into virtual nothingness by now. I do not know if this argument against BBT is valid or not, nor do I know if such an argument would also work against CUT to a much lesser degree. I simply mention it as another possible problem of BBT highlighted by a trace-back-to-zero scenario.

So, CUT absolutely rejects BBT's trace-back to a singularity's zero size. We didn't need an inexplicable singularity to bring us into existence. Our universe came from a real, comprehensible cause—something like a SUSE. Unlike a singularity theory, Cosmic Unity doesn't have to hide under the bed when questions relating to the laws of physics in our beginning universe come knocking at the door. Unlike the case with BBT disciples, no CUT believer will be told to discard reason and simply accept answers relating to a SUSE on faith. CUT celebrates the fact that the laws of physics that are said *not* to apply to the singularity and its Big Bang *do* apply to the SUSE that brought our universe into existence.

F. Increasing rate of expansion. Many BBT proponents and others are discombobulated by the fact that the universe is not only not slowing its rate of expansion, but the rate of expansion is actually *increasing*. From another factor that begins as a point of agreement between BBT and CUT (ie., that our universe is expanding) we see that another major divergence develops because of conclusions drawn from such fact.

The fact of increased rate of expansion appears to be an impossibility under both the basic open and closed BBT models of our universe. Under a cyclical closed model we should, some 14 billion years after the Big Bang, be at least beginning to slow down, preparatory to a possible later stop, followed by a possible later contraction that would develop into the Big Crunch, which in turn would be preparatory to another Big Bang. Yet the opposite seems to be happening. Similarly, under an open universe model of BBT there's no proven reason for our universe's expansion rate to be increasing.

It appears that there are at least several possible explanations for the increased-rate-of-expansion phenomenon. One involves a rather exotic, surprising, and extremely fine-tuned rationale. That explanation, of course, is the one that has to be selected by Big Bang theorists. It is presented by Big Bangers, for example, that our universe might have something like a "cosmological constant"[16] that could, once more, save BBT from demise. Again, we have to be tricky here. Even though we're talking about a *physical* dynamic, the cosmological constant-like force proposed by BBT proponents has the coincidental, serendipitous effect of turning on and off at just the right time to save BBT from what had been recognized as its rather recently-discovered "accelerated expansion problem." (Shades of *inflation* flexibility that is so helpful to BBT in coming up with answers for BBT's "horizon/smoothness problem," its "old stars" problem, and its "developed structures" problem.)

BBT proponents have suggested that their newly-reinvented cosmological constant is the explanation for the increasing rate of expansion recently observed to be occurring in our universe. The cosmological constant is said to be a force that supposedly has a normal positive gravity and pressure; however, about 7 billion years into the age of our universe, the cosmological constant switched over to an opposite, net negative, force. This changed force is now said to accelerate our universe's rate of expansion. BBT had no answer to the fact our universe was found to be increasing its rate of expansion instead of slowing down until this totally unforeseen cosmological constant phenomenon popped up. Whew! Just in the nick of time, too. No telling what would have happened if Einstein's original force (which he named the "cosmological constant") hadn't been converted to the side of the angels exactly when needed by BBT. The reason proposed for the change from positive to negative is that gravity's normal attraction in the universe was supposedly conquered by an outward repulsive pressure of a cosmological constant type of force. Big Bangers say that this change happened at precisely the right time in the history of our universe to explain why our universe is now accelerating, and not decelerating, as previously predicted.

I'm very skeptical of such fanciful arguments based on something not observed anywhere else, especially when it has the incredibly opportune effect of achieving another last-minute life-saving rescue for BBT.

"Dark energy" is another, presumably similar if not identical, proposed savior for BBT's increased-rate-of-expansion dilemma. Like the suggested cosmological constant, dark energy (sometimes referred to as vacuum energy) is a kind of force that, magically, can't be seen or explained compositionally, but, according to BBT proponents, just happens to supply a hypothetical push inside our universe or outward pull in space that supposedly explains the accelerating expansion of our universe.

As one of many examples, we can look at what professor Lisa Randall has to say about vacuum energy:

Even in the absence of particles, the universe can carry energy known as vacuum energy. According to general relativity, this energy has a physical consequence: it stretches or shrinks space. Positive vacuum energy accelerates the expansion of the universe, while negative energy makes it collapse.

..... The supernova measurements and the detailed observations of relic photons created during the Big Bang tell us that the expansion of the universe is accelerating, which is evidence that the vacuum energy has a small positive value.[17]

There are at least a couple reasons I don't believe in these BBT-saving theories. The first reason is simple recurring suspicion. If one doesn't get just a little wary when mysterious forces are pulled out of thin air or non-air (literally) time and again to save interpretations of a particular theory, perhaps one should remember the Brooklyn Bridge story. This accelerated expansion problem for BBT and its rather recent remarkable suggested solution forces us to ask again: why is it that so many totally foreign and speculative concepts are necessary to sustain the conclusions people draw from BBT in an otherwise reasonable world?

I do think that some BBT proponents are beginning to make the same conceptual mistake regarding their theory that the increased rate of expansion is due to dark energy as they did in the trace-back-to-zero process in arriving at a singularity, only in reverse. Since our universe has recently been found to be accelerating its rate of expansion, certain scientists now take the position that our universe will end, not in a crunch or as a frozen cinder, but in an unavoidable Big Rip. Because our universe is increasing its expansion rate they seem to automatically assume the absolute extreme: there's no stopping it. The galaxies, stars, planets, etc. will all be ripped apart by this mysterious overwhelming outward force. Again, I don't believe it. Just because our universe is expanding doesn't mean it can be traced back to no size; just because there could be another universe or verses doesn't necessarily mean that there is an *infinite* number of universes with infinite possibilities in them; and just because our universe's rate of expansion is increasing doesn't mean that such expansion will result in our universe's inevitable R.I.P. by rip.

In any event, though, IF some manner of mysterious force is acting to speed up expansion under BBT it would also probably act to speed up the expansion of our universe under CUT. Thus CUT would survive as an alternative theory to BBT, regardless (though BBT advocates might then very well inform us how totally *unreasonable* it would be for such a mysterious force to apply to a non-BBT-related-phenomenon such as to the remnants of a SUSE. Just think how unscientific and illogical that would be.).

But I mentioned that there was more than one possible explanation for the increasing rate of expansion phenomenon. The aforementioned cosmological constant/ dark energy/vacuum energy-type theory is the first. One of the main reasons I don't believe in this first explanation is that I think my second (and even third) possible explanations—which arise under CUT—are better.

CUT's chief explanation for an increased expansion rate does not involve strange forces of nature that switch on and off at what some might consider arbitrary times that just happen to coincide with what is deemed necessary to save BBT from collapse as a model for the origin of our universe. This explanation for the increased rate of expansion as suggested by Cosmic Unity is, as usual, very reasonable and very simple. All we have to do is put an eye to a telescope in our own universe and see dramatic evidence why cosmic objects speed up in their travels through space. Then we apply that observational evidence to the broad, unified, ordered scale of cosmic unity to be found in CUT.

I believe the reason for the accelerated expansion that we observe in our universe today is none other than gravity. (Einstein pointed out that the curvature of spacetime results in that force which we refer to as gravity.)

As we are well aware, as objects in our universe approach each other their gravitational fields begin interacting and their speed increases. Their speeds increase in proportion to their respective masses and distances.[18] As a very humdrum application of the principal, we know that our scientists use this cosmic fact of life to achieve "slingshot" speeds we would not otherwise be able to attain in sending space probes to other planets in our solar system. We see this principle in numerous ways on a daily basis in our universe, and some of our most spectacular cosmic views are pictures of two galaxies being pulled into each other at fantastic speeds generated by the gravity of their masses in space.

So, how, specifically, does CUT apply to the expansion phenomenon? My suggestion is that, after expanding for billions of years[19] after the SUSE originating event of our universe, our expanding SUSE remnants (in the form of our universe) have begun to interact in a noticeable way with the gravity force fields of large-scale structures that are located beyond the limits of our universe—at least our visible universe. Just as supernova remnants, planetary nebulae, galaxies, etc. in our universe speed up as they approach the mass of other celestial objects in our universe, so also is the same thing happening on a larger scale as the masses in our universe have encountered gravity from large-scale structure(s) in our surrounding parent universe.

Although such surrounding-universe structures are undoubtedly outside the horizon of our universe's approximate 14-billion-light year visible "mile-markers," and the effects of the gravity we observe from them may still be relatively weak, they are (and were) out there, and their gravity has probably been affecting our universe's rate of expansion to some extent for quite awhile. As objects in our universe and those in our large, all-embracing, surrounding universe become ever nearer to each other their speeds will of course accelerate accordingly.

So, our universe and its inhabitants—all composed of the remnants of the SUSE—are now being influenced gravitationally by the many surrounding structures of our parent universe. One can envision—and in fact one fully expects that there is—an incredible amount of cosmic mass that is beginning to gravitationally affect our universe. (One can expect that we would recognize most of the forms of such mass, and possibly not recognize other forms of same.)

That, then, is why our rate of expansion is accelerating. It's the gravitational effects from as-yet unseen mass in the surrounding medium of our parent universe. As the mass of our universe expands to rejoin its surrounding parent universe it is being gravitationally attracted in the same way cosmic bodies (viz., supernova remnants) in our universe are constantly being gravitationally attracted by other cosmic mass in the surrounding medium of our universe.

Such explanation for the increased rate of expansion of our universe is simple and wholly consistent with our Cosmic Unity Theory, and it's also wholly consistent with what we observe as ongoing processes in our universe, even on a relatively local scale. It's really a common occurrence, and such gravitational interactions should be expected if CUT is correct.

I suppose I could concede that the cause of the accelerated expansion of our universe is in fact an invisible, dark energy if everybody else conceded that the energy in question has already been named. It's called gravity.

In the mid-1970s scientists began noticing that our universe was not expanding uniformly. We're not referring to the increased rate of expansion discussed earlier, but rather the non-uniform direction and speed of expansion. Something "out there" was causing the Milky Way, our Local Group of galaxies, and even the supercluster to which the Milky Way belongs (the Virgo Supercluster) to be directionally pulled at a speed significantly greater than observed in our universe as a whole. There was quite a hullabaloo raised about it among astronomers, and in 1986 the direction of the "something out there" was tentatively identified, and the thing was dubbed the Great Attractor. The Great Attractor was hard to study, though, because it was aligned with the dust and gas along the plane of our Milky Way galaxy on the opposite side of the galaxy.

It wasn't until later, when astronomers from the University of Hawaii studied the area in question (referred to as the Clusters In the Zone of Avoidance (CIZA) study) by X-ray analysis, that it was discovered that the source of the attractive mass was much farther away than the Great Attractor region itself—at least several times farther away than the Great Attractor. It is now believed that the mass causing the attraction in question is coming from the mass of the largest concentration of galaxies in the observable universe, an area generally referred to as the Shapley Supercluster. There is, though, some question whether the source of the gravitational attraction has truly been identified. There is also the belief that large-scale "attractors" exist in other parts of our universe. My question: Are the sources of these large masses that cause such powerful expansions in differing directions all located inside our universe?

Under CUT the answer would be a probable no. Although the particular gravitational source studied in CIZA is probably inside our universe, the mother of all Great Attractors is our parent universe and the objects within it, hence the increased rate of expansion of our universe as a whole. Under CUT it is a logical and simple incremental step to envision that such influences on masses in our universe would be mirrored within the larger cosmos. We need look no further than Cosmic Unity to find the answer to the question relating to the increased rate of expansion now measured in our universe.

There is a third possible explanation under CUT for the increased rate of expansion of our universe, and it represents a slight variation on the last suggested explanation. One almost hates to mention it because some may feel that, by mentioning it, CUT adopts as an integral part of its theory the concept that our surrounding cosmos itself is in an expansion mode. CUT doesn't necessarily assume such fact. But that possibility does exist, and if true, a second alternate explanation for increased universe expansion under CUT is this: The surrounding parent universe itself may be in an expansion mode (for whatever reason), and as a result, the increased rate of expansion we now observe in our universe is part and parcel of such general expansion of the outside surrounding universe.

Thus the increased rate of expansion in our universe is most logically related to the gravitational effects of the mass in our surrounding universe—not some strange one-time-only theory based on a force never before observed and certainly never proven.

•••••

We might keep in mind that CUT's above-mentioned alternate explanations for the increased rate of expansion of our universe do not *have* to be the case in order for CUT to prevail over BBT on the origin-of-universe question. They simply offer reasonable alternatives regarding the observed accelerated expansion—ones that are consistent with Cosmic Unity, and ones that do not involve another of BBT's habitual rabbit-out-of-a-hat solutions.

Why should we stretch for strange, unnatural theories regarding the increased-rate-of-expansion question when we have sensible solutions before us for all to see?

CUT provides possible solutions based on logic and simplicity and observed reality.

G. Semantics: The pinpoint that brought our universe into existence didn't really "explode." This is not a major problem for BBT, but illustrates the difficulties one encounters when trying to harmonize things with a theory that needs so many unnatural accommodations to preserve itself. After BBT was introduced people at first freely referred to what happened at the Big Bang as some type of explosion. Such reference seems altogether reasonable, as one would naturally refer to anything else that went BANG in the way that was necessary to create our universe as an explosion. But BBT political correctness police began nixing use of the word "explosion" in reference to the Big Bang. They wanted to squelch two rather reasonable concepts relating to "explosions" when talking about the Big Bang. First, if one believes something was there to explode in the commonly-accepted sense of the word, it contradicts other aspects of the Big Bang model, (ie., the claim that there was no thing, no time and no place for an explosion of an entity to occur).[20] Result: These Big Bangers are left with claiming that their "Bang" was just a real, real fast expansion of an almost-nothingness that didn't exist in place or time. Secondly, in an explosion things fly away from the source of the explosion. But to allow such things to be hurled away from the point of a real explosion might appear to contradict the BBT concept that "space" itself is expanding (versus things in it), and a true-blue Big Banger can't have that. The result is that one just doesn't use the word "explode" in polite BBT company when referring to the Big Bang.

So, watch your language when talking about the Big Bang.

But CUT doesn't need to parse words here. There is no question that large stars can explode into supernovae, and there is no problem at all in stating that something as gigantic as a supernova occurring in our large surrounding universe went BANG in a huge way. Thus under CUT the bang from our SUSE was a big *explosion*. Things were hurled *away* from the explosion.

Semantic problems resolved.

H. Too much missing matter. We've previously commented on the fact that, unless about 90% more matter is found in our universe than we now know about, an oscillating, or cyclical, Big Bang/Big Crunch universe evidently just won't happen. Not only that, many scientists feel that something more is needed to hold our galaxies and universe together than just the gravity that would be furnished by atoms presently known to make up our universe. But some BBT advocates have come up with another construct to save their day. It's called "dark matter." The mysterious missing stuff exists, they claim, in a vast cache or caches that nobody has ever seen. One model (now the accepted Standard Model on the question) proposes that only about 4% of our universe is made up of atoms, 21% of dark *matter* and 75% of dark *energy*. Many scientists believe in dark matter, pointing to what they claim are its gravitational effects. Perhaps a slightly lesser number of scientists also believe in dark energy, pointing to (a) the fact that dark energy would precisely fill in the missing factor needed to provide enough gravity to hold things together in our universe, and (b) the increased rate of expansion observed in our universe. Regarding (a), in 2005 a British-led group of astronomers announced it had discovered the first "dark galaxy"——a huge, rotating, amount of mass with no stars, located about 50 million light years away.

But some scientists don't believe in dark matter or dark energy.

I keep a very open, though healthily skeptical, mind on the question of dark entities. But no theory of CUT has to depend in any way for its existence upon the discovery of unseen, unproven "dark" theories, especially dark energy.

Under Cosmic Unity the perceived "missing matter" puzzle may be solved. In calculating the entirety of mass in our universe nobody has counted the balance of matter, including the remaining core, of the SUSE that brought our universe into being. Additionally, the enormous masses and energy located in the rest our parent universe may provide a goldmine of missing matter that has never been taken into account by those concerned about missing matter.

I. Miscellaneous BBT problems. People find a host of other problems with the Standard Model BBT explanation of our universe's origin and post-BBT expansion. I've read through many of them in addition to those discussed in this book. Some of these suggested problems may be valid and others may not and, admittedly, a few are too technical for at least some of us cattle ranchers.

I again hasten to point out that, despite the adverse comments made here relating to BBT, this is not primarily an anti-BBT screed. CUT is in agreement with most of BBT's basic precepts. Playing off some of the numerous BBT failures and impossibilities, though, gives one the opportunity to show the favorable alternatives offered by a theory based on Cosmic Unity. Whereas certain conclusions relating to BBT seem more and more tenuous as time goes by, and evermore dependent on speculative exceptions to the usual rules of the game, I believe that conclusions relating to CUT will, on the contrary, seem stronger and stronger as time goes by.

I'm not aware of any possible valid arguments against BBT that couldn't be solved if we looked instead to a SUSE as the origin of our universe.

J. In the end, everything dies? BBT kills us all in the end. This is not a precept of BBT in itself. It's inherent in BBT, though. It's not even a "problem" in the sense that it is an impossibility or inconsistency, such as mentioned in the list of BBT's impossibilities or inconsistencies in the preceding pages. It's just disconcerting. For some reason it just doesn't seem quite appropriate to me that the ultimate future of everything everywhere should have to be certain death, possibly coupled with the termination of time itself (whether once or cyclically).

As we have discussed, one conclusion from the BBT model is that if there is enough mass in the universe to stop expansion and to cause a universal contraction, our fate is sealed, and that fate is a compressed death. Under such a "closed universe" everything is nothing more than one more page in the continuing book of cyclical Big Bangs. Nothing and no one—no matter how advanced and intelligent—survives outside the ultimate recurring Big Crunch that is the precursor to the next Big Bang. Under BBT's closed model, one can't exist outside the one and only universe, and periodically that whole universe is compressed into nothingness. Thus under a closed universe we (and any other life in the universe) must disappear into the squashed speck of a singularity.

As we have also discussed, if there is not enough matter in our universe to stop our expansion—and at present this seems to be the case—an equally dead-end scenario is presented under another version of BBT. Our ending under such an "open universe" is just lonelier and colder. We simply expand and expand and expand—all the while drawing farther and farther from other parts of our universe until, one by one, each galaxy, star, planet, moon and other celestial object becomes a cold, dark, dead, decaying piece of matter drifting forever into oblivion. (I'm not sure this absolutely has to be the case precisely, but that's the way matters are usually presented under the "open universe" hypothesis today.)

The more-recently proposed "Big Rip" ending for our universe also, of course, results in inevitable death of our universe, this time because of a suggested terrible tearing apart of everything as a consequence of a proposed uncontrollably-increased rate of expansion of our universe.

I don't wish to sound Pollyannaish about things, but the above catastrophic fates do not have to befall us under a Unity theory of the cosmos. Under CUT our future is unbounded (literally and figuratively). As time goes by we simply expand farther out into the all-encompassing next larger portion of the total universe (and/or there will undoubtedly be expansion into *our* part of the total universe of other matter, energy, and/or beings from such surrounding universe). Without question, such interaction between our universe and our parent universe has been taking place to at least some extent beyond the horizon for millions of years—and we may not be able to prove it for many years to come. There are all sorts of possibilities for us in a world in which our universe is a part of something larger. We (our descendants) will definitely have a future under CUT. Not so under BBT (or Big Rip or the many mega/multiverse theories that in essence subscribe to BBT's concepts of the eschaton in multiplied forms similar to open or closed universe endings).

Call me sentimental, but I like my theory better.

•••••

Although most of the foregoing problems attributable to BBT have been pointed out by various individuals who disagree with one or more aspects of BBT, we have yet to be presented a reasonable alternative theory to the contrived solutions Big Bangers offer for these problems. Under CUT, though, we not only see the numerous substantial problems confronting BBT disappear, but we're presented with reasonable alternatives in their stead.

What other theory offers all that?

5
Big Bang Evidence Proves Cosmic Unity Theory

As we have just seen, BBT's problems are multiple and varied. We have observed, though, how CUT shrugs off BBTs problems with hardly a sideways glance, with no need for exotic, fine-tuned solutions that challenge common sense. However, let's now look at the other side of the coin. Let's look at the bundle of evidence offered *for* the Big Bang as the origin of our universe. BBT proponents point out that the major proofs upon which they rely are not mere speculative musings but rather are based on actual measurements and mathematical calculations. They are thus claimed to be *observational* proofs for the beginning of our universe.

But in looking at such observational evidence for BBT one will immediately see that such concepts so enthusiastically advanced for BBT in point of fact actually constitute further *strong evidence in favor of a big bang occurring under CUT*. The result is that under CUT we can retain the cherished observational proofs for BBT as they apply to a big bang while at the same time discarding BBT's concomitant impossibilities.

Although the following pages state the main proofs for BBT, one should keep in mind that even much of this "observational" evidence for BBT has been challenged by a small but determined percentage of individuals who absolutely refuse to accept BBT for one reason or another.

Proof One: Background radiation. The cosmic microwave background (CMB) radiation that was accidentally discovered in 1964 (same year I graduated from law school, incidentally) by Nobel Prize winners Arno Penzias and Robert Wilson is said to prove that our universe began with a "hot" Big Bang. The bang is believed to have occurred about 14 billion years ago. The fact that cosmic background radiation was discovered to be isotropic (the same in all directions) and consistent with what scientists call a "blackbody" spectrum of approximately 3 Kelvin tipped the scales in favor of BBT for many scientists. The existence of such microwave radiation that apparently bathes our entire known universe was confirmed both by the Cosmic Background Explorer (COBE) space shuttle in 1992, and also more recently by NASA's Wilkinson Microwave Anisotropy Probe (WMAP). Even the measurement of small irregularities in such background radiation is believed to corroborate the fact that our universe's beginning was the Big Bang.

But although such left-over cosmic background radiation has been said to confirm *the* Big Bang, in reality it just confirms *a* big bang. As we have seen and shall continue to see, the theory with the best and widest-ranging big bang credentials and the least problems is one involving a CUT cause. Very possibly our SUSE.

I leave it to the mathematicians to calculate the heat that would be generated by the appropriate-sized SUSE that could have brought us into existence, but I for one have no doubt at all that it can be proven that a much larger-sized object than the Big Bang's singularity could have created the heat that has now dissipated (maybe over a somewhat different period of time than allocated post-Big Bang) to the 2.725 Kelvin we now observe extant in our universe. After all, we are aware of the tremendous energy and heat created by many supernovae in our universe. In many instances a supernova in our universe will outshine its entire home galaxy of a hundred billion stars for awhile. Then consider the fact that supernovae in our universe would be absolutely miniscule in comparison with monster parent universe star explosions. Just try to imagine the energy that would be generated by such a supernova in the next larger universe. We'll later comment on the comparative heat-generating incidents under BBT and CUT, but for now we can simply note that the cosmic background radiation taken as proof for the Big Bang also constitutes proof for the radiation resulting from the gigantic SUSE that brought our universe into being.

A second observation relating to CMB that is said to help prove that the origin of our universe was the Big Bang is that the temperature of the galaxies farthest from us appears to be higher than the temperature we observe in our own galaxy. Since the far-off galaxies that we see are the galaxies as they existed billions of years ago—when our universe was younger and hotter—this difference in temperature supports the BBT claim that after an initial hot bang we have been cooling as we aged and expanded.

But of course any temperature differences observed in far-off galaxies also support the fact that a huge parent universe supernova occurred billions of years ago, and the remnants of that supernova in the form of our universe would also have undergone greatly-reducing temperatures as the millennia have passed. Scientists have observed, not-so-incidentally, that remnants from supernovae in our universe go through the same hot-to-cooling-period phases. A SUSE under CUT is, therefore, proven by the difference in the observed temperatures in question.

A third CMB-related hallmark that is said to support BBT appears to be its large-scale "smoothness" (uniformity of radiation) throughout our universe. Actually, the perceived smoothness of our universe has been used as both an argument for BBT and against it.[1] But Big Bangers contend that only BBT explains the fact that radiation in space seems to be isotropic (the same in all directions) and homogeneous (the same at all locations).

However, look at the relative smoothness, isotropy and homogeneity of the remnants of many of the supernova in our universe, picturing yourself as situated among the remnants of such a supernova hurling out and away from the source of the explosion. Under such circumstances and with such a view and perspective you could very well believe that your universe is isotropic and homogeneous. Then increase the scale of the supernova and its colossal array of remnants surrounding you by a factor of many billions. In other words, assume that you are situated among the extended and expanding remnants of a very large SUSE. Where would such an assumption put you? Well, I think it might very well place you exactly at our spot in our universe.

The expanding remnants making up the Crab Nebula have already, by way of a very tiny and local analogy, achieved a diameter of some 60 trillion miles in just the 950-plus years since the supernova explosion that created it. There are other supernova remnants within our galaxy that extend out for quadrillions of miles from the supernovae that caused them. And that's just within the immediacy of our Milky Way Galaxy. Now contemplate the apparent smoothness resulting from an immensely scaled-up supernova—a SUSE—that exploded with more than the condensed mass of the entirety of our one-hundred-billion-plus galaxy universe. I do believe that the apparent smoothness and homogeneity from the perspective of someone observing and measuring within such an immense, expanding, stretched-out "universe" of remnants and radiation would be exactly what we see and measure today.

The smoothness of the CMB is one of the arguments Big Bangers use in stating that the universe did not really start with a "bang" of any *thing*. We can see some irregularities in the CMB, but Big Bangers say that if our universe had begun with the explosion of some *thing* there would be many more, and more pronounced, irregularities than can now be seen. They observe little or no evidence of thinning out or bunching up of galaxies out near the edge of the 13-to-14 billion light year distance of our visible horizon, and they detect little difference in CMB radiation as they look in different directions. That may be true. But I would ask: (1) So far as observational evidence is concerned, what about the relative smoothness or irregularities of that much-larger portion of our universe that lies beyond the horizon? What does that look like now? Just maybe it would look a bit like the sometimes-more-jagged or frizzled (less-smooth in radiation) outermost areas of many supernovae we see in our universe. And: (2) So far as CMB detection is concerned: (a) One truly wonders just how much directional difference in CMB we could necessarily detect from the perspective and location of earth if our galaxy is in fact buried within the enormity of the remnants of a SUSE that occurred billions of years ago; and (b) One wonders what the CMB differences (if any) would be some 14 billion light years out past our horizon?

Maybe this CMB proof-positive for the Big Bang as the sole possibility for an originating mechanism origin for our universe isn't quite as strong as it's held up to be.

There are also these two questions that could be raised regarding whether CMB constitutes unimpeachable proof that our universe originated with the Big Bang:

(1) What would be the natural expected approximate minimum temperature of bodies in space in our universe, regardless of anything called the Big Bang? Some scientists argue, on several grounds, that the average temperature of space would not be absolute zero, and that it's wholly reasonable to believe that a temperature a couple degrees above that lowest possible number should be expected. These scientists say that a temperature of about 2.8K—which is the approximate temperature of our measured CMB—would in fact be the natural average minimum temperature of space in our universe.[2]

(2) What would be the average temperature in the space of our surrounding parent universe under CUT? Is there CMB that fills our total surrounding parent universe? An argument could be advanced that the existing temperature we observe in the space inside our universe might also be the natural average temperature in all of space. There is certainly a strong argument that, with all the anticipated cosmic activity in the surrounding parent medium, there should be some resulting radiation extant. Maybe the temperature of the surrounding universe in the vicinity of our universe averages about 3K or so, regardless of a long-past big bang of some kind.

Thus, there is at least the possibility that the measured temperature of CMB in our universe doesn't constitute proof of one particular cosmic event called the Big Bang as the only conceivable dynamic for the birth of our universe. It has also been pointed out by some scientists that the apparent isotropic nature of the CMB could be the result of absorption and re-emission of the radiation by the multitude of dust clouds and galaxies in the universe. In a fog on earth, for example, the whole sky might seem to have the same muffled brightness throughout, even though we know that the sun—on the other side of the fog—is located in one specific direction and place in the sky. Therefore, goes the argument, the CMB does not inherently exist in all places as a result of the Big Bang that began as a singularity.

Regardless of all of the foregoing, however, for purposes of this discussion I shall nonetheless proceed on the assumption that the 2.725K measurement we observe in our universe is the true attenuated result of a gigantic big bang billions of years ago....And what would be true for the bang's radiation history under BBT could by the same token be pretty much true for CUT when all adjustments and relevant factors are correctly considered.

So, despite questions about just how strong CMB really is as a proof for BBT, one can surely argue that, to the extent that it *is* such proof, it works for a SUSE under CUT just as well. In fact it works better for a SUSE than for a beginning non-physical thing that suddenly began to expand from nowhere at a rate probably faster than the speed of light for some unexplained reason and for what could be considered an arbitrary period of time.

The final conclusion on this point, then, is that the several particulars of the pro-BBT argument as they relate to CMB only help firm up a SUSE position. If we grant that our universe did start with a hot bang that resulted in radiation that is measurable today, such start probably wasn't the inexplicably strange, hot bang of BBT.

It was the hot bang of our SUSE.

Proof Two. The expansion of the universe. When Einstein's work was reviewed by the very gifted Georges Lemaitre in the mid-1920s the conclusion reached by Lemaitre was that Einstein's calculations lead to an ever-expanding universe. Lemaitre, a Belgian priest, wrote a paper on the subject in 1927. He then met with Einstein and discussed his belief on this point. Einstein was favorably impressed with Lemaitre's mathematics, but he disagreed strongly with the idea of an expanding universe.

Both men, who had each arguably started out on the right foot regarding the expansion question, resultantly later traveled down apparently wrong paths.

Einstein, who, despite the findings of his mathematics, felt that our universe is in a steady state, came up with a formula that provided an unknown force to supply just enough anti-gravity pressure for the universe to remain in what he believed was a steady state. His formula provided a "cosmological constant" for our universe. After Edwin Hubble's later proof that our universe is expanding, Einstein withdrew his cosmological constant thesis, believing it to be perhaps the biggest mistake he ever made.

For his part, Lemaitre proceeded on the assumption that, since the universe each day was larger than it was the day before, its expansion could be traced backward and backward and backward. Ultimately one reaches an essential speck of nothingness. This unbelievably tiny beginning point was later referred to by Lemaitre as the "primeval atom," and is now referred to by the scientific community as the "singularity" that later exploded into the Big Bang.

If Lemaitre and Einstein, who remained on friendly terms over the years, had only been able to reach collaboration on expansion maybe they would have come up with a trace-back to reality of an expanding universe instead of Lemaitre's trace-back to zero that was later embraced by the scientific community as a whole. In the opinion of this writer, the myriad problems facing BBT today began with the trace-back-to-zero process itself, and have been resolutely compounded ever since. Of course there is the possibility that any such meeting of minds just wasn't very likely until Hubble, and later Arno Penzias and Robert Wilson, made their respective discoveries relating to expansion of our universe and to the cosmic background radiation lingering in it.

In any event, as a result of Edwin Hubble's work we know that our universe's clusters of galaxies are all expanding away from each other in all directions. Hubble calculated that the velocity that a galaxy recedes from us (measured as a red shift—toward the longer wavelengths in the electromagnetic spectrum) is in direct proportion to its distance from us (Hubble's Law). Thus the farther away a galaxy is from us the faster it is receding from us. The balloon analogy has often been used to show how things that start at a common "location" can expand away from each other in all directions, just as our universe is believed to be doing after the Big Bang.

Today almost everyone agrees that our universe is expanding.

But BBT's "expansion" proof applies equally well to our SUSE model as to the Big Bang model.

As already mentioned, probably the biggest error of the Standard Model BBT is an implied assumption that, since our universe is expanding, at one time it was compacted into an incredibly dense, tiny speck so small as to be immeasurable. Once that trace-back-to-zero path is taken other errors are sure to arise if the concept of the beginning tiny singularity is not abandoned. Time and time again we see the expression by Big Bangers of such a trace-back assumption., and I won't bore the reader with quotations from the experts on this point. Basically, every book, lecture, radio or television program today subscribes to the inherent assumption that, since our universe is and has been expanding, there was no point at which it could have begun its expansion except at about zero size.

Reference to logic, simplicity and experience should not be out of line here. BBT's procedure of calculation backward into nothingness is a process that would be untrue if applied to *anything else* that has expanded—whether an atomic bomb, a hand grenade, a firecracker, a kernel of popcorn, or a star in our universe that exploded into a supernova.

BBT supporters simply continue to insist on applying concepts that are applied to nothing else. But why *should* we go along with a contention that our familiar universe should be the lone exception to the rule that things must come from something? And from what we can see there is always at least *some* size relationship as things evolve, grow, or separate one from the other. The beginning of our universe, for some perverse reason, is thought by Big Bangers to be a wild exception to the laws of physics and reason in this regard also. (The explosion of a compressed star in our universe into a mini-world of supernova remnants, for example, does not even come remotely close to what we're asked to believe regarding the infinitesimally small dot from which the entirely of our universe allegedly grew.) But acceptance of such an outlandish process is akin to an article of religious faith that some BBT disciples might ridicule in other contexts.

To me, it's much more reasonable to start with a presumption that our universe is NOT an exception to logic and experience. To me it's much more reasonable to trace our beginnings back to something that actually existed in time, was possessed of physical size, and had a beginning—just like everything else. Since the universe that sprang from our beginning is immense, just maybe the object from which it sprang was fairly large itself—especially in comparison to the pinpoint of nothingness envisioned under BBT. I'm not saying that it wouldn't be theoretically possible for a universe to begin with some sort of otherwise-unheard-of process involving, for example, interaction with a tiny piece of negative matter that itself came from nowhere, but why reach for pure guesswork when we have experience and reason as an alternative?

Once more we point out that a SUSE is our perfect candidate representing experience and reason. (Say, maybe we'll find that size really does count—as when starting universes.)

I have to repeat: *Just because something is found to be expanding doesn't mean that it didn't have a measurable size when it began expanding.*

•••••

So, in view of the obvious expansion of our universe, what should scientists do in calculating backward to the beginning of our universe? They should simply calculate the compression of our universe back to a real object—to the size, density and composition of the appropriate parent universe star that collapsed/exploded into a supernova, resulting in the remnants of which we are the product. We know the answer. We are the answer. Let's work back to the correct question under a theory of Cosmic Unity.

Calculation of a projected temperature and size of our universe in its initial stage(s) would be integral in such a process. I've seen various estimates regarding what the temperatures of our universe would have been when our universe was of different projected sizes as it is thought to have expanded from the singularity. I don't know if they're right or wrong, but let's jump into the size of our universe at some early stage of development. For example, one projection is that when our universe under BBT (1) was compacted to the approximate equivalence of the density of the nucleus of an atom it (2) would at that point have had a temperature of a trillion degrees,[3] and (3) would consist of a sphere with a diameter of some 10,000 astronomical units.[4]

Whether these figures are close to reality, the foregoing is the *process* in which I believe we should engage under CUT in order to establish the early conditions of our universe based on an assumption that our universe developed from a real object possessing real size (ie., a SUSE)—as opposed to automatically assuming that we must trace back the size of our beginning universe to zero. As stated, I am hypothecating the above-mentioned example from someone else, but if we use figures like those above-mentioned, what adjustments would have to be made to begin our universe with a SUSE in order to arrive at where our expanded universe is today? Without "inflation"? Without other BBT cheat factors? I would think that such calculations would be both challenging and immensely exhilarating. My only caution for someone engaging in any such calculations is that, in arriving at a proposed answer, he or she not resort to any special BBT-type accommodations just to come up with a solution consistent with Cosmic Unity.

I believe, though, that we'll find that it's much more reasonable that our universe would have begun from an actual, relatively familiar, object with comparatively large size that truly exploded in time and space, than from a totally alien entity of no size that did not explode, did not exist in time or space, and is said to have expanded many times faster than the speed of light for an arbitrary period of time.

In closing on this point relating to BBT's expansion-related-proof we must recall that the expansion of our universe as presently conceived under BBT still requires a rate that is arguably impossibly fast at its very early "inflationary" stage. Under CUT, though, the expansion rate from an object incredibly larger than BBT's object of zero size would not have had to be impossibly great or the result of concepts that apply nowhere else. Mathematicians, as noted, can start doing the calculations.

Proof Three. The age of the universe. Scientists believe that the age of our universe is about 14 billion years. This has been determined through several methodologies. Examples of age calculations that are said to confirm BBT: (1) the *rate* of expansion of the universe as scientists have measured it; (2) the estimated age of the oldest observed stars in our galaxy and the age of the oldest galaxies outside the Milky Way; (3) the measured rate of decay of certain radioactive elements, followed by backward calculations in time at their known rate of decay; and (4) the age of various globular clusters whose age we can reasonably calculate.

Depending on who is doing the calculating, all of the foregoing are said to prove that the age of our universe is about 14 billion years old, which is stated to coincide with BBT. But again, it must be pointed out that all these age measurements could just as properly prove that a SUSE occurred a requisite period of time ago. Nothing in these age calculations excludes our SUSE candidate. And if we end up with a somewhat different age result under CUT I am sure we'll all find a way to adjust and live with any such cosmological addendum.

Although the age of our universe is claimed to constitute a proof for BBT, in truth the age argument (like the smoothness "proof") is also used against BBT, with some contending that the ages of certain galaxies and large-scale structures in our visible universe are too old to have begun just 13 to 14 billion years ago. This is sometimes referred to as

BBT's "large structures" problem. Some experts contend that the true age of our universe would have to be as much as 100 billion years—not BBT's approximate 14 billion years—to account for the developed large structures found in our universe.

A calculation of the age of the universe under CUT—without BBT's magical shortening of time during which the universe is said to have expanded faster than the speed of light—would seem much more in keeping with reality. If BBT's inflation falls as a theory, the stated age of the universe may increase. For our purposes here, though, we'll simply assume that the age proof works as a proof for BBT; but at the same time we'll also assume that BBT's age arguments would also apply to a SUSE.

I can't help repeating, though, that since the universe that sprang from our beginning is immense, just maybe the object from which it sprang was reasonably large itself—especially when the suggested alternative is the pinpoint of nothingness envisioned under BBT.

Again, a SUSE is our perfect candidate for such a reasonable beginning.

There are at least two considerations that might change the age estimates of our universe if it arose under Cosmic Unity as opposed to BBT. These two considerations are in conflict. On the one hand, the comparatively large size of our beginning universe with a SUSE under CUT would probably militate in favor of a younger age for our universe under CUT than under BBT, other things being equal. This would seem reasonable in that a beginning size many orders of magnitude larger than a singularity of zero size should—again, other things being about equal—require less expansion time to reach the present size of our universe under CUT than under BBT.

On the other hand, though, we must keep in mind that "other things" *aren't* equal. In its age calculations BBT employs a timed physical miracle (inflation) that decreases the age of our universe by speeding up its expansion rate to velocities much faster than the speed of light. That aspect of BBT, of course, militates in favor of a much younger age for our universe under BBT than CUT (not considering for a moment the fact that our universe started at a much smaller size under BBT). Also, of course, any actual differences in expansion velocity would obviously be a component to factor in.

If we throw out BBT's inflation cheat factor what would the *net* effect be on the age of the universe? Our universe would probably be older, which would then help explain some of BBT's problems relating to the old, developed structures in the universe, the horizon problem, etc. But we'll let the numbers people counterbalance the two dynamics and come up with a reasonably accurate answer. No problem there. Just another little math calculation, right? It's my feeling that we have more wiggle room in being able to adjust the commonly-conceived *age* of the universe than we have regarding certain other so-called pillars of proof for BBT. Therefore, as calculations are made relating to the beginning of our universe from a SUSE instead of a singularity I think that there will undoubtedly be at least some change in our belief that the age of our universe is exactly 13.7 billion years.

Proof Four: The high incidence of light elements. Scientists observe that we have a much higher percentage of light elements in our universe than one would expect to see except for the occurrence of the hot Big Bang. They point to the presence of natural deuterium, as well as lithium and lithium-7 and helium and helium-3 and helium-4 as examples of otherwise-unexplainable high concentrations of light elements.[5] Predictions that nucleosynthesis resulting from the Big Bang should have resulted in a percentage of about 75% hydrogen to 25% helium are said to be proven observationally in our universe. It's believed by BBT supporters that the high incidence of light elements in our universe can *only* be explained by virtue of our universe's beginning as a superheated mass of ultra-hot protons and neutrons that subsequently cooled—which is exactly what they feel happened with the Big Bang. This position is possibly weakened to some extent, though, by some studies that show that all but the very lightest elements we see in our universe could in fact have resulted from processes that stars undergo (ie., in supernovae). In 1957 three British citizens and an American published the famous 108-page "B^{2}FH" (2 Burbidges (Geoffrey and Margaret), William Fowler and Fred Hoyle) study that showed, among other things, how nucleosynthesis takes place in stars. This publication is one of the most widely-cited studies in this general field of science.

Regardless of whether the Big Bang proponents are or are not right in respect to their claim that only the very highest temperatures could have produced the light elements in question, the fact remains that, if they are correct, it could have been a SUSE that provided those needed high temperatures. And if our universe did start with a SUSE instead of the singularity's Big Bang we would still be very well situated to possess the requisite percentage of heavy and light elements we see in our universe today. I can think of several bases for such a position, and I'm confident that supernova specialists could probably do a much better job of addressing the matter than I.

Several thoughts come to mind regarding this question of light element percentage in our beginning universe:

First, it is probably true that our universe started with a high percentage of light elements, but it may not have been in the *precise* percentages envisioned under BBT. Depending on the true age of the universe and its beginning size, beginning heat and density, etc., under CUT there may be opportunity for some variation of the exact beginning percentages of light elements.

Second, our universe did not necessarily *have* to begin with a basically all-hydrogen/helium mix from an entity that is said to have originated from a nothingness that didn't exist in place or time; that depends for its existence on (arbitrary?) early inflationary adjustments; that depends for its existence on unreasonable speculations requiring increasing rates of universe expansion; that depends for its existence on numerous time-and age-related contradictions; and that didn't really explode; etc.

What if the entity back to which the scientists work their calculations turns out to be, say, a huge (bigger than anything we imagine in our universe), short-lived (relative to the parent universe) first generation parent universe progenitor star that was extremely high in light elements? The explosion of such a parent universe star could have resulted in a scattering of existing light elements inherited from *its* progenitor star in much the same way as we believe occurs in our universe after stars go supernova. The remnants of that explosion could in turn have produced stars, or subsequent generations of stars, with the right percentage of ingredients for the creation of the star system in which we now reside. Maybe the heavy elements we see in solar system are, for example, still a product of three generations, but two are from our universe and one from our parent universe. There are a number of possibilities here.

Third, even if our universe did start with the 75%-25% hydrogen mix in question (or close to it) there is the probability under CUT that our SUSE actually *produced* (didn't just scatter inherited) light elements.

Question here: Did the SUSE that originated our universe have enough heat to create the light elements?

Well, as we noted previously, if it's heat we're looking for as a basic part of the process that created our light elements, a monster supernova occurring in our next larger universe should be one of the most intense heat sources imaginable. Consider that a supernova in our universe with a star mass of only 8 suns is said to attain a temperature of over 10 billion degrees less than one second before exploding. What would be the heat generated by the explosion of an incredibly gigantic parent universe star that contained the equivalent mass of more than *ten sextillion star systems* of our universe, plus all the additional matter in our universe's intergalactic medium? It's hard to believe that the heat generated by a SUSE that contained at least the mass of our universe wouldn't be sufficient to create conditions that would produce the light elements under discussion.

And as noted several paragraphs above, studies have shown that the processes involved in supernovae in our universe can produce virtually all the naturally-occurring elements, including light elements.[6] If in fact light elements can be created by any non-big bang processes in our universe the "light elements" pillar of proof for BBT would be considerably weakened and the position of CUT solidified even more.

In any event, a SUSE possessed of the mass of our universe (and probably much more) would certainly appear to be possessed of the heat necessary for the production of light elements. And if, as a part of the light-element-forming process that later devolves into generational star formation in our universe we need a cooling process to follow the "big heat," we merely have to look at the second phase—aptly named the "cooling phase"—of a classical supernova life cycle in our universe to get what we're looking for on that point.

So, if in a SUSE we have the heat and cooling mechanisms necessary for creation of light elements and their later progression into heavier elements through generational star evolution, why couldn't those same light elements touted as having come from "the" Big Bang just as easily have resulted from the bang of our SUSE?

In our SUSE scenario what we're talking about is *at least* the same amount of mass that the Big Bangers conceive of at the singularity, but it is squeezed together as a result of a parent-universe-progenitor star collapse and supernova explosion instead of (a) a singularity caused by an oscillating-universe-Big Crunch and subsequent Big Bang, or (b) first-time singularity that becomes the Big Bang.

Admittedly, in BBT the mass is thought to be crunched into density a pinpoint in size versus the SUSE's much larger (and arguably much less dense) crunch size. This difference in density would, though, be lessened to some degree by the fact that the SUSE had an original total mass (which would include a very substantial as-yet unseen probable remaining core in some form) that was in fact considerably larger than the combined mass of what we consider to be our universe. Whatever the comparative aspects of the two hypothetical masses and their respective densities at boom time, I have a strong feeling that scientists would be able to confirm that the SUSE we're talking about would create an unfathomably enormous amount of heat, wholly sufficient to result in a hot plasma soup leading to the subsequent production of the light elements in question.

But let's say a skeptic of CUT would next contend that if a prospective SUSE contained the mass of our entire universe it would simply collapse into a huge black hole and never explode outward because of the immense gravity of the concentrated mass.

I would point out that such a question would be a very rational one to pose in respect to a Big Crunch or beginning singularity scenario. No adequate explanation has been offered for how the singularity's escape-from-gravity occurred. As observed previously, though, supernovae don't have the understandable problem in this area that the singularity/Big Bang does. We *know* the physical processes that take place when a big star collapses and is then blasted out into space as supernova remnants. CUT, therefore, has an observed, known answer to the gravity problem. BBT doesn't.

One might point out in further defense against a gravity-related argument directed at CUT that such an argument doesn't and didn't prevent BBT proponents from subscribing to a cyclical or singular Big Bang theory despite the fact that they had to invent, again, a new theory (a "bounce") to accommodate it. So, if the gravity problems don't dissuade scientists from BBT, they surely shouldn't dissuade scientists from a SUSE under CUT.

Of course it's more than fair to ask: If CUT is right, what was left behind in our parent universe after the explosion of the SUSE? Something akin to a black hole? A neutron star? Pulsar? White dwarf, etc.? Good question. Well, in all probability there is some kind of remaining object. It's outside our visible horizon right now, but someday those who follow us will surely know what it is. (Note that we don't take the position of certain BBT believers who say that if we can't see and measure it it's not a profitable thing to think about.)

It's very probably out here. You can guess what it will look like.

This "high incidence of light elements" proof, then, may simply be additional corroboration that *a* big bang occurred, and not that it was *the* Big Bang so widely championed. After all is said and done, our SUSE is the perfect candidate for the bang that brought us the high incidence of light elements presently found in our universe—whether the light elements were created by the SUSE or merely scattered and delivered to us via its remnants.

I have every confidence that a SUSE occurring billions of years ago has yielded the appropriate percentages of light/heavy elements that presently exist in our universe.

Other Proofs. The above-listed "big four" proofs are often referred to as the very pillars that support BBT. There are other lesser-discussed arguments occasionally offered in support of BBT from time to time, but none that I am aware of can exclude the hypothesis that the big bang that really started the whole thing was a SUSE.

I take the fact of the expansion of our universe as BBT's strongest "pillar," with the existence of an all-embracing microwave background radiation as a close second (many scientists would rank CMB number 1). Third strongest is the high concentration of light elements argument. If further studies show that light elements can be created in the processes of normal star evolution or other processes, this BBT pillar may be weakened considerably. The remaining main pillar of BBT proof—the age of the universe—which is calculated to be about 14 billion years, works for CUT as well as BBT, but in my opinion the estimated age of our universe probably has room for refinement. And as mentioned earlier, it is this pillar that may have to be adjusted to some extent if a SUSE becomes recognized as the standard model for the origin of our universe. Once (a) BBT's inflation cheat factor is taken out of the age equation, and (b) the larger beginning size of CUT's SUSE is added into the equation, the calculated age of our universe could be adjusted accordingly. Things will nicely fall into place, and it shouldn't be a forced fit, as is the case with BBT.

Thus all of the "big four" pillars of BBT stand at least as strongly for CUT as for BBT; and each argument, proof, or supporting piece of evidence that is offered for the favored BBT merely adds to the viability of a SUSE model that is unencumbered with all the Big Bang impossibilities and inadequacies noted in Chapter 4.

6
What Kind Of SUSE Brought Us Into Being?

We're free-wheeling now. Once we accept the propositions in the foregoing chapters the rest is simply speculation (and the foregoing *wasn't?*).

If in fact our universe originated with a SUSE of some sort one of the questions that comes to mind would be what *kind* of supernova or collision in our next larger universe brought us into being? Could our universe have evolved from the remnants of, say, just any old type of parent supernova or collision of celestial bodies?

No, probably not. Applying the anthropic principle,[1] one could say that our causative SUSE was the very kind, and perhaps the only kind, that could have resulted in what our universe is. Otherwise we wouldn't be here to talk about it.

However, if we want to move past that sometimes-accepted concept of why we exist today, there are a number of possible star-related cataclysms that come to mind. The "E" in "SUSE" could just as easily stand for the broader "*Event*" as for the possibly more limited "*Explosion*." For purposes of this discussion, though, I am going to continue to concentrate my remarks on supernovae, pointing out that in just this general category of star activity alone the possibilities for our universe origination are many. Even our relative primitive knowledge about supernovae in our universe allows us to divide supernovae into various classes and sub-classes; and from what we have been able to glean thus far there appears to be much we still have to learn.

A few examples of the tremendous variety of factors we observe in supernova, even with our comparatively recent studies of supernovae:

- The spectra found in Type I supernovae in our universe do not contain prominent hydrogen lines, whereas the spectra found in Type II supernovae do.
- Some supernova remnants are of uniform density and consistency, while most are not.
- And while some supernovae in our universe do not produce measurable amounts of hydrogen, helium, oxygen, carbon, silicon, neon, magnesium, sulfur, iron, etc., others do.

- After the initial explosion and shock wave of a supernova in our universe a cooling takes place, but the cooling rate differs from supernova to supernova, and in fact the rate, or stage, of cooling can be different *even within the remnants of the very same supernova*.
- The several phases of supernova remnant evolution in our universe differ greatly in length and speed. In the cooling phase of supernova remnants in our universe, for example, the period of cooling can differ by as much as a thousandfold—from a mere 100 years up to 100,000 years.
- Remnants of supernovae in our universe are usually divided into three or four general types, each consisting of about four phases, with a wide variety within each type, sub-class of type and phase. Some of the better known supernova remnants just within our tiny neighborhood in our universe are the Veil Nebula, the Crab Nebula, Puppis and Vela, but it's noteworthy that even among that small sampling of four we see different type-classifications for supernovae.
- What is the *generation* of a particular star that went supernova? It is the generation of the exploding star that determines, for example, many of the differing chemical characteristics of its remnants mentioned above. As the star generations progress away from the time of the supernova explosion the chemical offspring of such supernova explosions contain a higher and higher concentration of the heavier elements. Supernova experts believe, for example, that our solar system may be "third generation" because of the existence of heavy elements that would not exist in an earlier generation star explosion.
- We also know that the very *location* of a star that goes supernova in our universe is important insofar as the ultimate creation of new stars, planets, moons, asteroids, etc. is concerned. When a supernova occurs near a molecular gas cloud a veritable nursery for new stars can develop. The shock front of the supernova accelerates, mixes, compresses and heats the gas, and can change the chemistry of the gas, thus creating the conditions for new star formation.

The last-mentioned possibility may be exactly what happened when our progenitor star in our surrounding parent universe went supernova. Its remnants could have been flung into a large gas cloud of the surrounding medium, with its shock front creating the requisite environment for our resulting universe.

We can see the remarkable star-generating effects of supernova explosions near molecular gas clouds in several locations in our universe as we view them today.

•••••

The whole point of the above listing of supernova variety and conditions is this: With such a huge diversity of possibilities flowing from supernovae in just our universe alone one can scarcely even begin to guess at the enormous window for SUSEs and supernova diversity within the immensity of our large surrounding universe. Supernovae might occur on the order of billions of times per second within the enormity of our parent universe. And of course much of what we have said regarding a parent universe supernova might also apply to, say, a parent universe planetary nebula and large-scale star collisions and explosions of various types.

Thus our universe most reasonably came from a cause outside our universe, and that cause was probably an exploded star of one kind or another that occurred in the existing space of our physically-surrounding cosmos. Within the incredible array of possibilities that must exist in such surrounding medium there would certainly be found a plethora of candidate progenitor stars or occurrences that could have created the remnants from which our universe was born. As we've also mentioned, we're dealing with an immense difference in scale when we try to compare things in our universe to our huge surrounding universe. Scale models usually don't track properly when trying to analogize objects with size differences ranging into large orders of magnitude. In the parent-child relationship we're considering here the disparities in volumes, masses, and dimensions could be so great that many analogies might be inappropriate even in view of application of the same basic laws of physics.

That having been said, though, when considering SUSE candidates one naturally thinks in terms of the progenitor star to which we owe our existence as having at least some of the qualities and composition that would be understandable to us. One wonders: what generation was the progenitor star in our parent universe that brought our universe into being? Because of the high incidence of light elements that apparently existed early in our universe was our progenitor star the equivalent of what we would consider a first generation star in our universe? It must be stated again that the explosions, implosions, collisions, mergers, and interactions of gigantic cosmic entities in the immense surrounding universe existence may not track exactly with the smaller ones with which we are familiar in our universe, even with the same general laws of physics. We'll speculate about this further in the next chapter but we'll have to leave it to the passage of time and our ever-developing expertise in cosmology to come up with more definite answers in attempting to identify a more specific Type classification of parent universe supernova that probably brought our universe into being.

What if a question is raised against CUT to the following effect: If our universe is the result of a SUSE or some type of surrounding universe nebula, *where is* the remaining core mentioned as a possibility in the last chapter? The object that would be the equivalent of our universe's expected black hole, neutron star, pulsar, white dwarf, etc.?[2] Simple, and we have already commented on this. If the parent universe explosion was of the type that left some sort of residual core—and it probably was—such remaining entity is still physically out there in the surrounding medium in some form.

No problem there. We just can't see it yet.

As we are well aware, at present we find it very difficult to see through to much of what is on the other side of our own 100,000-light-year-wide Milky Way galaxy. In fact it hasn't been all that long ago that we discovered that there may well be a supermassive black hole in the middle of our Milky Way galaxy, and we aren't even sure we can actually observe the details of its visible void yet. So, one shouldn't be surprised that we have not yet detected structures lying billions of light years away, especially those lying out past the visible horizon. It would also be fair to keep in mind that a sizeable separation has undoubtedly developed between the point of original SUSE explosion and its remnants (in the form of our universe) during the past billions of years.

Once we develop better instrumentation—away from earth's atmosphere—we will of course be able to better inform ourselves about the nature of our cosmos; but we must always remember, though, that the big limitation to observations past the approximate 14 billion light year mark has nothing to do with either instrumentation or the interference caused by earth's atmosphere. Rather we're limited by what we can see in the duration that light has had the time to reach us. We will never be able to contemporaneously see to the very edge of our universe, or even to many less distant celestial objects, unless we can one day somehow overcome the limitations imposed by the speed of light.

Despite the fact that we cannot see out past the boundaries of our universe I have the feeling that we will someday be able to come up with some pretty good parameters for determining the physical characteristics of the specific entity in our surrounding universe that caused our universe to be born.

Just how comparable would supernovae in our universe be to large star supernovae in the parent universe of which we form a part is, as noted, unknown, and there are all sorts of concomitant questions that arise when thinking about the matter. The differences in what happens at the death of various-size stars in just our universe alone is astonishing. The informational payoff in comparing details of supernovae in our universe with details of our parent universe's supernovae would be fantastic. We would certainly gain unimagined insights into the nature of both worlds. Cosmologist Timothy Ferris states that cosmology differs from other sciences in one important way. Other sciences have a variety of things they can use for "comparisons." "Cosmologists," he points out, "have only one universe to study."[3]

Well, if Cosmic Unity holds up as a theory, such deficiency will be corrected. Think of all that we may be able to learn by respective comparisons between our universe and a next larger surrounding level of universe that probably possesses the same basic laws of physics that we have. (In fact we already have one small idea of the instructive things that can result when comparing supernovae in just our universe to our own universe at large. The teams of astronomers who made the startling discovery that our universe's rate of expansion was accelerating, not decelerating, were actually studying expansion rates of supernova remnants in our universe at the time.) In decades and centuries to come the possibilities for comparative analysis and extrapolation between supernovae in our universe and related cataclysms in our surrounding universe will be truly absorbing. This, in fact, is one of the major reasons why I wholeheartedly disagree with those who say that working on, or thinking about, things past our visible horizon "is not a profitable thing to think about."

Frankly, I don't know or really care into which category the exploded progenitor star of our universe is ultimately found to fit. As discussed, it may well be a classification we haven't yet delineated, and of course it may be one that doesn't necessarily precisely correspond with supernovae in our universe. My main thrust here, as always, is that our universe originated with *some* causative event emanating in our surrounding universe, and CUT holds that that cause was most probably a SUSE or planetary nebula of some sort. At the same time I hold open the possibility that the cause might have been any one of a variety of actions in our parent universe—all the way from an intentional act by intelligent being(s) to something that was solely an occurrence of nature in such surrounding medium.

For now I'm opting for an act of nature in the form of a SUSE.

7
More About the Physical Connection Between Our Universe and the Surrounding Universe

With our four basic propositions discussed in Chapter 3 in place let's now take things a few steps farther. (More of the speculation that I had originally intended on avoiding as much as possible.)

Let's continue to assume we came from a SUSE, and continue to assume that the laws of physics between our universe and our parent universe are mostly compatible: Exactly what is the actual *structural relationship* between us and the next larger universe? How are we morphologically attached?

I've touched on this to some extent, and there will be some repetition here. My suggestions, followed by some random thoughts:

Our universe is totally surrounded by our parent universe. There are none of the tunnels, passageways, umbilical cords, walls separating islands, black hole entrances or exits, trapdoors, bubbles, or membranes, etc. that are sometimes pictured as either connecting cosmic worlds or keeping them apart. As previously mentioned, our universe can be thought of as the area that constitutes the *extended blast field* of a huge star explosion that occurred billions of years ago within the pre-existing space of the universe that physically surrounds our universe. Such blast field is occupied by the remnants of the SUSE, together with the spaces and voids that have evolved within such field in the ensuing years following the giant explosion.

As we have commented on at length, such a concept is of course diametrically opposed to the Big Bang theory that conceives of nothing *outside* our expanding universe, and *nothing preceding our universe into which* we are expanding.

Just as an explosion on earth (and everywhere else we've ever heard of) creates a temporary microcosm of intense energy and expansion as things are launched away from the point of the explosion, the same thing has happened following the explosion that resulted in our universe coming into existence. Of course it has happened on a scale appropriate to the gigantic parent universe star involved. The detonation of an atomic explosion here on earth results not only in blast damage as the force expands, but also as things are sucked back into the vacuum created by the blast. Obviously the latter phenomenon does not happen in a cosmic explosion, as a virtual vacuum already exists in space, which is to say exterior to the blast field. The result in space, as we see with supernovae in our universe, is that the remnants of the blast just keep expanding farther out into space, sometimes evolving into other structures, sometimes eventually coming under the gravitational influence of other cosmic mass.

I would not, by the bye, anticipate that our universe is a perfect sphere (or perfectly flat or saddle-shaped, for that matter).

Below is a two-dimensional drawing showing how CUT conceives that our universe would be located in the vastness of the physically-surrounding cosmos. Because of scale considerations no other parent uni-

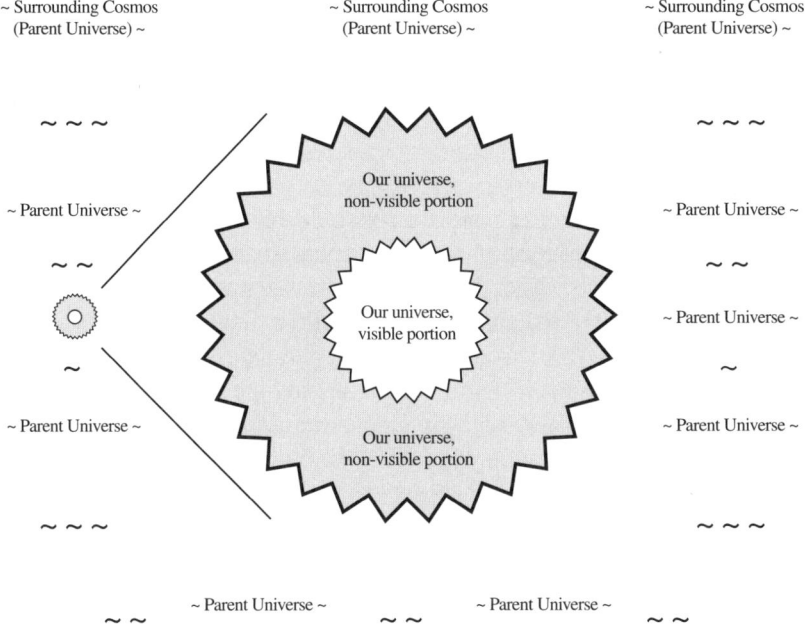

ENLARGED VIEW OF OUR UNIVERSE

Graphic. The visible portion of our universe is surrounded by the non-visible, or unobservable, portion of our universe. Outside the entirety of our portion of the total universe (which we refer to as our universe) is the finite space of the surrounding total "parent" universe of which we form a small part.

verse bodies are depicted in such graphic, though many decillions of same would be found there.

Under CUT our universe would certainly not be virgin from contact with the "outside" surrounding medium. As our universe has expanded in all directions it has continuously come into contact with outside particles, cosmic pressures, energies, and gravitational influences of varying intensities. There is also the very real possibility that some celestial bodies or particles from the surrounding medium have been hurling toward our universe from the outside and making indentations into our universe during the past billions of years. It's unimaginable that there has been no entry at all into the now-much-dissipated and greatly-extended blast field of our SUSE by material or energy from the parent universe. (Could any of the neutrinos that are passing through our very bodies today have originated in that part of the surrounding cosmos outside our universe?) We might also keep in mind, regarding the shape question, that the farthest points of our universe in all directions and at each millimeter of separation will all not be the exact same distance away from the originating SUSE location, because the energy and various particles of matter have not all been traveling away, unimpeded, at the same speed along every point. Thus, for these and other reasons, one should not anticipate that our universe would be perfectly even, or smooth, physically at its outermost boundaries, or possessed of a perfectly symmetrical shape.

More speculation: Let's briefly consider the size of our progenitor star. Why was the progenitor star that exploded in our large surrounding universe so much bigger than the stars we see in our universe? There could be a number of reasons. I tend to believe that in our surrounding parent universe there is a fantastic variety in sizes of all objects, including stars. There may even be varieties in sizes of entities that are quite foreign to us, even though subject to the same physical rules of nature. Regarding stars, though, the fact of the matter is that some stars are probably immense; many others are smaller, and many more still are *comparatively* tiny objects. Our progenitor parent universe star was obviously a member of a class of stars that would seem gargantuan to us— a class that can contain masses larger than our entire universe. I don't know why there can be such a variety in size of stars in the larger, surrounding, universe. Maybe it's just the natural course of things. It may be the ultimate fate of most of the matter in our universe and the surrounding universe to eventually be gathered and accreted into the large objects, including stars, that exist in such surrounding medium. (Oh, no, not another Big Crunch scenario.)

·····

    Just because we speak in terms of a "large," or parent, or total, universe does not mean that everything adrift in it is larger. Here on earth we have, for example, a huge difference in the size of animals found in our oceans, lakes, rivers and streams. A blue whale, a denizen of our large oceans, can weigh in at 120 tons, tens of millions of times the mass of individual members of some of the small krill species. And smaller still are various insects and other life in the form of, say, bacteria. For every large whale, porpoise, shark, sea turtle, etc. existing in our surrounding oceans there live billions of smaller creatures and members of lower life forms. We may have a somewhat analogous situation involving our universe and our huge surrounding universe. There could be large numbers of extremely big structures in the next larger universe, but these numbers might be dwarfed by all the smaller bodies also extant there. Maybe the things found in our universe are part of the large array of the many smaller structures that exist throughout our parent universe.

    Carrying the ocean analogy a bit further, one could point out that, even though our oceans and small creeks exist in a planet possessed of the same basic laws of physics, many animals that live in one body of water could not survive in the other body of water, for various reasons. For one thing, most fresh water fish cannot long survive in salt water. Is there something in either our universe or our surrounding universe that, because of differences in scale or mere accidents (not fundamentals) of nature, acts to affect the size and the character of objects within them over a period of time? Even though they possess the same basic rules of nature? I don't know.

    One can muse that such diversity in size of objects located in our next larger universe makes our universe vulnerable to being captured by the gravity of the big fish swimming in that vast surrounding ocean. (And incidental to that, won't such a possibility make things interesting for our descendents in millennia to come? But then such things will occur in the slow motion applicable to such large entities and enormous distances, won't they?) As we have already discussed, CUT suggests that the very fact that our universe's rate of expansion is increasing could indicate that our universe has probably already caught the gravitational eye of some very big body(ies) swimming in the waters of our surrounding universe.

    Obviously, entire treatises could be written on the morphological possibilities extant in a next larger, surrounding, cohesive universe.

    I have no problem visualizing that our universe is possessed of the physics that could also accommodate the incredibly large stars of our parent universe; it's just that none has developed within our visible universe to date because there hasn't yet been sufficient passage of time within the blast field of the SUSE since our remnants were blasted into existence. As noted above, maybe everything we can see (including our solar system and every thing and every being in it) is at this very moment involved in the long-term process of evolving into bigger objects that can only presently be found in our parent universe. (And I'm not talking about the obesity problem confronting many humans today.) In time, possibly the stars, galaxies and galaxy clusters in our universe will eventually form into larger and larger aggregations of matter until, after a very long while, they become part of one of the leviathans of our parent universe. Possibly the voids that we presently see between these concentrations of matter in our universe will expand in proportion to the continuing process of enlargement of celestial bodies being formed outside the voids. Eventually these ever-more-concentrated collections of matter may evolve into parent universe single objects in somewhat the same way as big objects in the form of planets, their satellites, stars and galaxies developed from the initial primordial mass of our young universe.

Our solar system and galaxy could, of course, be just one of many trillions of microcosms of such a process. A very local example: Astronomers feel that at one time in the evolution of our solar system some 20 "planets" had built up, plus, obviously, many, many other smaller bodies. These larger bodies continued the process of crashing into each other and merging until we now have eight (poor Pluto) major planets and a number of planetary satellites and assorted minor planets, comets, asteroids, ice dwarfs, debris, etc. Some of the moons will eventually be drawn into the planets they orbit; some moons will drift and crash into each other; some planets and moons will be affected by our sun's red giant stage, etc. On a much larger, extra-solar level, galaxies and galaxy clusters will continue to merge. (No one knows how many smaller collections of stars our Milky Way galaxy has attracted to itself in this process over the eons.) And of course we know that our huge Milky Way galaxy is headed into a probable collision and eventual merger with Andromeda.

The above speculations could have full application in a CUT world, but not in a BBT version of an open end universe in which things would expand forever into cold, dead, decaying emptiness.

Even if the ultimate fate of most matter in our universe is *not* continual concentration into larger celestial bodies it seems obvious to me that at least a fair portion of the matter comprising our expanding universe will find itself in time being gravitationally attracted to, and linking up with, sibling structures of some kind in our surrounding universe. The interaction between objects in our universe and the surrounding medium will be, and undoubtedly already is, ongoing. It's all in the nature of Cosmic Unity.

Black holes and their role in the parent-child universe phenomenon will probably remain a question for quite some time. It may be that black holes, just like almost everything else in our universe, will get swallowed up into the huge stars or other objects that reside in our parent universe. For example, in our universe we observe that black holes—and there are evidently many billions of them—are the apparent big bad boys in their localities. They can swallow up much that comes within capture distance of their formidable presence. But what if these black holes, even the so-called super-massive black holes, run up against objects billions or even trillions of times larger than they are? They haven't had to face up to such prospects in our visible universe yet. They may be in for a very big surprise.

On earth we see many whirlpools in small bodies of water, where they seem to be the masters of their local stream, creek or even an occasional river. But do we see them in vast bodies of water? Do we see them controlling their environment in the expanse and pressure of the world's big oceans? No. The same with tornadoes. They may be powerful masters of their limited local environments, but they're hardly a drop in the bucket when it comes to our planet's atmosphere as a whole. So perhaps when a child universe's black holes come up against cosmic bodies that absolutely dwarf their size and mass (such as parent universe stars) they're gobbled up with the same ease that smaller objects in our universe are gobbled up that wander too close to stars or black holes or big planets. Black holes may be bullies in our universe, but nothing more than tasty morsels for gargantuan stars or other entities in our next larger universe. Interesting thought here: If a black hole in our universe with ten times the mass of our sun meets up with a parent universe star with ten quintillion times the mass of our sun, our insignificant little black hole isn't going to eat the monster star, now, is it? Wouldn't it to be the other way around? Whether the black hole would even cause the star indigestion might be the real question. Wouldn't it be intriguing, though, to watch that interaction?

Of course it could very well be that powerful black holes of large size do exist in our parent universe. The remaining "core" of our SUSE may be such a big black hole.

If big black holes do exist in our surrounding medium, another interesting proposition arises. What if the particular parent universe body (or one of them) causing the increased acceleration of expansion of our universe is a black hole? Odds are very probably against it, but what if we're in the somewhat early stages of being drawn toward the event horizon of a monster parent universe black hole? On such an off-chance the Big Rip people might, like a few others we've noted, be right for the wrong reasons. Time (a lot of it) will tell.

If we continue our assumption that our universe evolved from a SUSE, further interesting considerations arise. We've already discussed the fact that if our universe started with a parent universe supernova there are several possible explanations about how we can nonetheless arrive at the correct percentage of light elements extant in today's universe. We've already mentioned that the progenitor star that exploded could have either created the light elements we have observed or been the delivery system for scattering such light elements. In respect to the prior suggestion that the SUSE itself could have created the light elements, though, the idea occurs that *all* large-star supernovae in our immense surrounding universe might well result in production of only (or almost only) light elements. The reason for this is that the incredibly mammoth implosion/explosions inherent in such extraordinarily huge parent universe supernovae would practically always *have to* reduce everything into a superheated primordial soup in much the same way as envisioned by the singularity's Big Bang. Heavy elements simply wouldn't be able to develop or retain their more complex structures in such colossal detonations that would be so incredibly intense and productive of such unimaginable heat. In such cases, the Big Bang's theories relating to nucleosynthesis, production of light elements, and star evolution might be able to be salvaged and morphed into an explanation of what happens with the occurrence of almost *all* huge-star supernovae in our large surrounding universe.

Under the above-mentioned Cosmic Unity concept of our surrounding universe there might exist stars and supernovae in a variety of sizes, with a variety of evolutional generations and a resultant variety of light and heavy elements created from the star explosions. In other words, stars of differing "generations" would undoubtedly not be unique to that part of the total universe of which our universe forms a part. Differing star generations should exist in the surrounding cosmic medium also, and, once more, size might really count in what they produce when their life cycles end.

And this obviously brings up other considerations. I'm just rambling now (I don't have a professional career in cosmology to worry about), and these thoughts aren't tied directly to a belief in CUT.

Although there is always the possibility, and even probability, of exceptionally large terrestrial, or "rocky," planets in our parent universe, their number may actually be somewhat self-limited under the concept just mentioned above. The reason for this is that the later-generational heavy-element-producing stars that go supernova in our next larger portion of the universe probably *can't* be the biggest stars there. (As we noted a couple paragraphs above, the biggest stars that go supernova in our parent universe are so massive that they may only produce first-generation-type worlds that are rich in the light elements.) Consequently, the smaller parent universe stars—more in line with stars we see in our universe— that go supernova not only generate less heat and light elements than their monster big brothers, but they (a) throw out a lesser mass of material that (b) later develops into remnants that (c) in turn later evolve into heavy-element objects such as planets, moons, asteroids, ice comets, etc.

Thus it could be said that the very *size* of the stars in our next larger universe may put a cap on the amount of heavy or light elements that will be produced there. This isn't so much a consideration in our universe, as we probably don't have stars big enough to result in the sole production of light elements as a consequence of an out-of-this-world big bang. I'm not sure what the approximate maximum size a star could usually be when it goes supernova without creating a cataclysmic "start-over" big bang that produces only the light elements, but I'm guessing that the parameters of such a thing would not be all that difficult for the number guys and gals to calculate. Again, it's only math. Probably back-of-the-envelope stuff.

This leads to the further (wild) conjecture that parent universe life in general and parent universe beings in particular could possibly have more rocky planets with somewhat "reasonable" size and tolerable gravity—resulting in *somewhat* more familiar environments in which to exist—than might at first blush be expected in the immensity of the surrounding next larger universe.

There is also this possibility under CUT: CUT could be right about the origin of our universe, right about why it's evolving, right about why our universe's rate of acceleration is increasing, and right about how it is structurally related to a larger, surrounding parent universe, *but* maybe any other *our universes* within the parent universe have somewhat varying physical laws. Thus the fact that CUT is right regarding the origin and location of our universe and some related details doesn't necessarily carry a guaranty relating to the details of siblings in the parent universe.

Another thought: Under CUT the total surrounding cosmos may be static except for local disturbances. The atmosphere on planet Earth is static in the sense that it only moves around within the immediate vicinity of planet Earth. Our Milky Way Galaxy rotates, but is mostly static in that it does not appear to expand or contract in and of itself. Maybe our parent surrounding universe is static in the same way that Earth, the solar system and the Milky Way are static despite the occurrences of many explosions, collisions and cataclysms of various type that cause individual local expansions and contractions inside it. By this I do not mean to say that I'm adopting in any way SST's concept of continual creation of matter from nothing as it might apply to the large surrounding universe. Not at all. I'm just saying that our surrounding parent universe under CUT is not *necessarily* in an expansion mode, even though that relatively small part of it we are referring to as our universe is expanding within the framework provided by the surrounding medium. Maybe Einstein was right in respect to his original belief that the universe is static.[1] Maybe he was just was just one universe size off. Of course if our surrounding universe is neither expanding nor contracting, some of the same problems Einstein wrestled with in the early 1900s might arise again. (For example: What's to keep the larger surrounding universe from eventually imploding under its own weight?)

Besides, it's now mostly accepted that our universe has to be either expanding or contracting. But what about the total universe?

What fun.

•••••

Obviously the above-mentioned questions and speculations do not *have* to be the case in the large surrounding universe, but we're just tossing out possibilities right now. As I've said before, once we get past the basic propositions of Cosmic Unity mentioned in Chapter 3 and bolstered in Chapters 4 and 5 we're much freer to guess about the details of the SUSE and its world. Once we've opened the right door the rest can be quite engrossing. (I do not, as stated before, want such further speculations about the possible characteristics of our next larger universe to contaminate CUT itself. CUT certainly doesn't need to depend for its existence on the validity of my guesswork relating to any particular characteristics of our surrounding parent universe.)

As soon as we let the cat out of the cosmic bag and allow ourselves to consider our universe in terms of an expanding entity within a greater surrounding universe the additional follow-up questions that come to mind are practically endless.

Just a few to start things off:

What are the varieties of celestial bodies that could reside in a larger total universe with basically-compatible laws of physics?

Is our parent surrounding universe expanding, contracting, or in a steady state, and why?

What is the age of our next larger universe?

How big is it?

What is its shape?

What was the origin of our parent universe?

Even if our *fundamental* laws of physics are the same, how much could things still be different between the our universe and the surrounding cosmos?

Will such basic compatibility in our one, unified, total universe actually make communication between the parts thereof a greater possibility?

A probability?

Are there other sibling "our universes" created by other SUSEs from our surrounding universe?

Are they common or not?

Under CUT everything is supposed to be a part of something larger; so how does this apply to our surrounding universe?

How distant is, and what's the nature of, any remaining core of the SUSE that brought us into being?

How close are we to areas where our universe and our surrounding universe meet?

Have parts of the surrounding universe made incursions into what we consider to be our universe?

Substantial incursions?

What is the nature of any such incursions; how did they come about; and where did they occur, or where are they occurring?

What are the parent universe masses that could be causing things in our universe to accelerate their rate of expansion?

How far away are they, and what are the parameters of distances we're talking about in respect to all these matters?

What are the time elements that apply to our parent universe?

Is there an "outside" to our parent universe?

Are all the questions we ask about our universe and our parent universe operative for our parent universe's relationship with a still-larger universe, etc.?

How do these considerations increase the likelihood for the existence of life, and especially intelligent life, in the total surrounding universe?

Do these considerations increase or decrease the probability of a prime mover?

Has our universe been observed by, or even visited by, beings from our surrounding total cosmos?

If so, where and when?

Have any beings in our universe communicated in any way with beings in the surrounding universe?

In view of the reality of Cosmic Unity, how much of BBT should be scrapped and how much salvaged?

When will we have answers to some of these questions and many other questions inherent in the foregoing?

To address the last question: I don't think we must resolve the above questions in order to come up with an answer to the original inquiry of this book. As noted in the first sentence of the Introduction our mission was to determine the origin of our universe. *Our* universe. (If CUT were a theory that our universe is just an atom of a larger universe, which is only an atom in a larger universe, *ad infinitum*, we would not, of course, have come much nearer to a workable theory than we have under BBT. Such theory of strict verticality of universes, incidentally, is a result of what Joel Primack would call (and we have previously referred to as) Scale Confusion and Scale Chauvinism—beliefs rooted in thinking that what occurs on one size scale can automatically apply at much larger and smaller scales.[2] And it seldom, if ever, can.) But CUT is totally conceptually different from such an approach, even if CUT does envision that our universe lives in a large, surrounding and compatible universe.

Our universe is part of a unified cosmos that originated with an explosion in the next larger level of that cosmos—our physically-surrounding parent universe. A leading candidate for such explosion would be a parent universe star explosion, and our universe is composed of the expanding and evolving remnants of that explosion that occurred billions of years ago. The explosion itself could well have been a supernova occurring within the medium of such surrounding cosmos. All these concepts just don't fit into a theory that our universe is simply an atom-of-an-atom-of-an-atom-of-an-atom-of-an-atom.

Solving our original question shouldn't mean that we now immediately have to solve the questions that will naturally arise regarding the specific origin, history, and structure of our surrounding parent universe. I repeat what was said before: I really do think we should be entitled to approach this one universe, or part of a universe, at a time. We haven't yet seen our parent universe, and one can scarcely even conceive of it. I, for one, have a problem getting any kind of mental grip on what might be the immensity of the surrounding cosmos. Just imagine, our parent universe could well be as much larger percentage-wise to our universe as our universe is to, say, one of our universe's quadrillions of average stars that go supernova in one of our billions of average galaxies in one of our millions of average clusters of galaxies in one of our probable thousands of average superclusters of galaxies. A truly incomprehensible concept.

Thus, I'm not going to begin to speculate at length right now about the origin of the surrounding parent universe. I respectfully note that my purpose in this writing is to propose an answer to the Big Question relating to our universe. If someone complains that Cosmic Unity (1) only solves Big Bang's quandaries, and (2) only solves the questions of how and why *our* universe originated and how it's structured I'd say that that's all I set out to do. Such a complaint would be a bit like the man who searches his whole life for his father. When he finally has a DNA test taken and establishes who his father is with a 99.9% degree of certainty he rejects the test in disappointment. Why? Well, he now says that the test didn't tell him who his *grandfather* was.

The fact is, we humans accept many, many things, and are absolutely sure that they are always true, without knowing all the reasons or origins or details for same. Perhaps Cosmic Unity now tells us for the first time what we are and where we came from and where we're situated and where our descendents will go. Isn't that what we've been wondering about all these years?

•••••

Let's address multiverse theory directly.

For various reasons CUT is not a multiverse theory. To begin with, CUT envisages our universe as a congruent part of one, total universe with basically-compatible laws of physics—not simply one of a patchwork of separate universes with varying laws of physics. As we have noted throughout this book, it's a hallmark of multiverse theory that separate "verses," as a matter of definition, necessarily equate into separate laws of physics.

Also, under CUT our universe did not violate the law of conservation of matter by self-generating, or self-creating, itself as we see in most multiverse models. Quite the contrary, it's a basic tenet of CUT that our universe results from a pre-existing star-like entity that physically exploded. Our universe did not come from nothing.

Another distinguishing factor between CUT and multiverse: Unlike most multiverse theory, CUT does not claim to be part of a mysterious chain of eternity, with no ending, no beginning, and with such an infinity of real and artificial possibilities that nothing is known or knowable. CUT takes a diametrically opposing view: Under CUT our universe is a part of a physically-surrounding total cosmos that is very real and finite.

Also keep in mind that as an adjunct of most multiverse theories we of course are told that we'll never be able to communicate with entities outside our universe, not only (a) because of differing physical laws, but also (b) because these multi, mega, bubble, perpetual, or pocket universes are thought to be flying away from each other at fantastic speeds. We've already addressed (a) above—CUT just disagrees with the premise of (a), as explained throughout the entirety of this book. Regarding (b) above, though: CUT disagrees with (b) because under the CUT model it is fully expected that the surrounding universe and its contents will one day impinge on and enter our portion of such total universe. Simply put, under CUT our universe will totally physically mix it up with our surroundings, and it has undoubtedly already done so to date. Swapping spit between the two is part and parcel of the CUT model morphologically. Our universe will not forever expand away from all other things any more than supernova remnants in the Milky Way Galaxy will expand away from everything else forever.

Furthermore, unlike various multiverse theories, it's not a foregone conclusion under CUT that expansion applies everywhere. Whether the all-encompassing total universe that surrounds our universe is necessarily all expanding away from itself, either as a whole or in our locality, is at least an open question.

Additionally, as mentioned in Chapter 4, BBT kills us all in the end. And the same holds true for most multiverse theories, which are based, as we have seen, on the open-universe premise that such bubble, perpetual, or pocket, worlds are all traveling away from each other forever at such enormous speeds that they'll never be able to catch one another. Such multiverse premises condemn their ever-expanding bubbles or pockets to the same cold, lonely, dead conclusions as BBT. One of the exceptions is the megaverse concept of death by way of the alternate BBT-type closed universe in which the brane-type universes implode in a big crunch every trillion years.

But CUT, unlike BBT or the various multi/megaverse theories, holds out the sure prospect of absorption and continued assimilation of our universe into the surrounding life of the total cosmos.

Most multi/megaverse theories don't resolve the various expansion, horizon, age-related and other problems that exist with BBT. Simply saying that there are multiple copies of Big Bang creations doesn't in and of itself solve such BBT problems (although admittedly some would be solved if the CUT approach to the individual "verses" were adopted). And still further regarding the multis: What the heck is that material usually graphically depicted as surrounding the various pocket, bubble, mega, perpetual, or multiple universes? Is it space? Whose space? What brand of space? Is it nonexistence? Etcetera. Under CUT we know what surrounds our universe. What surrounds the multiverses?

The foregoing, then, are a few of the major points of distinction between CUT and what is normally considered multiverse theory. Although throughout this book we devote much space to showing the differences between CUT and BBT, the truth of the matter is that some of the BBT claims and assumptions that we challenge are the very same claims and assumptions we would also challenge as made by multiverse proponents in their various duplicated and multiplied forms.

•••••

I don't know if our incredibly immense surrounding parent universe is a part of a yet-larger universe. As noted, it would be nice to see at least one parent universe structure outside our universe before being forced to walk the plank about the details of that entire surrounding universe. That wish aside, though, we can probably at least nibble at the edges of the parent-universe question. I will avoid the obvious temptation right now to hypothecate that the origin of our parent universe was the remnants of a supernova-type entity from *its* next larger universe, etc., although that, of course, is a possibility. It may be, though, that when we reach such immense scales we have exceeded the scope of humankind's present ability to relate or comprehend. I do feel it can be said that our original premise about everything being a part of something larger will definitely apply *in some way* to our surrounding universe. I'm not going to speculate any further on the size question right now.

Secondly, Cosmic Unity holds that the basic laws of nature in our next larger universe will not be contradictory to those in our universe once any differences resulting from scale would be taken into account. The most likely scenario in all things, including universes, is that a child will be like its parent in most material ways. This general principle should apply to our next larger surrounding universe and any cosmos surrounding it. As mentioned before, though, when we talk about such enormously large scale differences some reality may be quite perplexing, even under basically compatible laws of physics. There may be animals in that bigger zoo in the sky that would be difficult to grasp intellectually, regardless of beginning compatibility of laws of physics.

Then of course there's always the very reasonable possibility that there simply ain't no bigger nor more complex animal than the large, surrounding medium. Maybe there doesn't have to be.

Referring just to our parent (not any possible grandparent) universe now: I do not mean that things will inevitably look and act the same in the surrounding universe simply because of compatibility of physical laws. There will undoubtedly be many exotic things in our giant parent universe. And some *accidents* (versus fundamentals, or constants) of nature could vary from one parent universe structure to the next, thus possibly causing enormous apparent differences from one to the other. Further, as observed before, even though the same basic laws of science apply, it may be that something occurring on a tremendously larger scale than we comprehend has results not imaginable to us in our universe (For example, undreamed-of out-of-this-world "tipping points" of nature may be reached based on large scale difference alone). Such differences may even result in what amounts to total phase transitions of the things in question (Ie., as when, because of differing temperatures, ice turns to water or water turns to steam).

But the elements (including those elements as yet undiscovered) will remain the elements, will remain the elements, will remain the elements, even though they might sometimes manifest themselves in ways alien to us. So we should expect that water would boil and water would freeze in the larger total cosmos under the same conditions (if and when we could find them there) as exist in our universe.

I do believe there is a reasonable possibility that some entities in our next larger universe may not be all that outlandish to us. Even aside from the compatibility that one would expect to naturally result between the remnants from a supernova and its progenitor star (and even such star's companions in the total universe), other considerations enter in. If we address the possibility of life forms in our parent universe, for example: The very laws of physics in a unified cosmos suggest that life forms, and certainly complex life forms, probably conform to at least some basic requirements. Things simply can't grow and sustain themselves if they don't have *some* necessary means of fueling and re-fueling themselves, eliminating or converting wastes, conforming to gravity, and experiencing some sort of sensation and interaction with their environment, etc. Through the processes of evolution, probably only the most efficient such forms will survive, just as we believe happens on planet earth.

By the foregoing comments I certainly do not mean to exclude the possibility of alternate biochemistry, either in our universe or the larger cosmos. But even this could exist under basically-compatible laws of physics.

We have some very strange-looking creatures here on earth, especially in the deep sea, in caves, and in multifarious other odd places. We may be surprised at how unsurprising some life forms in a parent universe would be in comparison with earth's life forms. As just one of what could be many examples, assume we would encounter in our parent universe a larger-than-Jupiter-size rocky planet with a very large gravity field. (We could, of course, even encounter such planets in our own universe.) Any beings living on the surface of such a planet would obviously not be able to look and act like us. But after the possible shock of seeing them we would at least comprehend *why* they look like they do. Life would thus not be surprising in the sense that it would be wholly unfathomable to us; it would be more a matter of possible revelation at how the life in question is presented in physical form to us under the circumstances of the particular environment.

The foregoing having been said, I do believe that (1) if and when we begin encountering things at greatly differing size scales almost all bets are off, and (2) recognizing certain alien life forms in other locales—at any scale—may well be a challenge, even with application of compatible laws of physics.[3]

I have dealt very little with theories of other dimensions, parallel universes, dark energy, string theory or M theory. Admittedly, I'm somewhat skeptical of certain of such concepts. They're much more complex and speculative than necessary (Ockham's Razor) for me to make the main points of this book. Maybe if I were more schooled in physics and mathematics I would be more understanding and supportive of some claims made under such theories. But since these notions aren't needed in order to understand the model of Cosmic Unity and how it applies to our universe and our surrounding parent universe, I abandon ruminations regarding such theories entirely to others.

It's fair to ask: If Cosmic Unity Theory, as opposed to Big Bang Theory, is cosmologically right, what difference will it make to us? (Aside from putting some people out of work for awhile—But then think of all the people who will suddenly find new work; think of all the books that will be written, all the lectures that the experts will give, all the new cosmological activity that will burst upon the scene.)

And even more significant potential arises, if that's possible. The very future of our universe, for example, suddenly changes. Our future becomes much brighter indeed.

Under Cosmic Unity we will someday be able to gain tremendous insight into our universe and its future from observations and comparisons between our universe and its parent universe. We will finally have another universe with which to compare ours and to make reasoned and reciprocal predictions.

And remember how pessimistic the experts were about us ever being able to communicate or interact with beings or life forms outside our universe? Such strongly-held beliefs, based on (1) the presumption of totally different physical laws in any world other than our universe, and (2) the supposed inability to ever catch up with other "verses" speeding away from us at impossibly great velocities, might well be swept away by the possibilities of Cosmic Unity. The goal of communication with beings in our surrounding universe may one day replace our existing goal of someday communicating with other beings in our own universe.

And what about the ultimate fate of our universe and the beings in it? With CUT we are no longer talking about BBT's (and most multiverses') certain, implacable, destruction and death of everything and everyone. Under Cosmic Unity our universe is open, with a continuing future of life for us and our descendents.

We couldn't ask for more.

And think of the albatrosses that CUT would allow to be flung from around the necks of BBT apologists.

People can now abandon the truly ridiculous claim that our universe came from *nothing*, and that no matter or energy preceded it.

People can now abandon the claim that the beginning entity of our universe, the singularity, did not exist in any *location* at all.

People can now abandon the claim that *time* did not exist before the bang that brought our universe into being.

People can now abandon the claim that our universe expanded (inflated) (a) *faster than the speed of light,* and (b) for a much-too-convenient while.

People can now abandon the claims that our universe has *no boundaries or edges*, and that there is no such thing as being inside, as opposed to being outside, our universe.

People can now at least think about abandoning or modifying the claim that our universe's rate of expansion is accelerating solely because of an unseen *hypothetical dark energy, vacuum energy, or a cosmological constant.*

People can now abandon the claim that our universe did not begin with an *explosion.*

People can now abandon the general claim that *galaxies don't really move and expand through space*, but rather that it is only the space itself that is stretching.

People can now abandon the continued attempts to come up with a theory explaining why the universe is smooth (assuming it is smooth) in view of the *horizon problem.*

People can now abandon the totally *arbitrary trace-back method* for our universe based on the assumption that if something is now expanding it must have begun with essentially zero size.

People can now abandon the idea that the condensed mass of our entire universe could possibly *overcome and expand away from the gravity* that would have existed in a singularity.

People can now abandon the position that any existence other than our universe must by definition be possessed of other physical laws.

People can now abandon the abandonment of hope that we can ever determine the beginning size and nature of our universe because of the concept that it began at Planck length, and all conventional concepts of space break down at such immeasurably small size in the inscrutable realm of quantum physics.

Some of the above considerations are interrelated, but the above catalog is just a start on a list of BBT claims that we can now cast into the (interesting) dustbin of cosmological history if Cosmic Unity prevails.

At the same time we should remember and incorporate into CUT the several viable points of BBT.

Under Cosmic Unity we can and should preserve the *general* concept of expansion. The electromagnetic redshift evidence for it remains compelling. It's true that Big Bangers don't admit that matter itself is moving out and away from the bang, claiming only that the fabric of space is swelling. BBT seems to be right, though, that a *type of expansion* is going on. CUT says it's the usual kind we see everywhere else; BBT says it's a unique fine-tuned breed of expansion. But a form of expansion definitely seems to be occurring.

A parenthetical note before proceeding to the next concept of BBT that can possibly be retained to some degree under CUT: It has to do with an aspect of BBT's concept of expansion. If CUT is right, then the Steady State theorists did stumble onto truth in one area of their expansionary-versus-static-universe controversy with BBT, while Big Bangers have in their turn stumbled into error in other aspects of their revered expansion proof for BBT. Under CUT we see that SST could be considered to be right in that the "state," or the size, of the location where the explosion that began our universe occurred has *held steady* after the explosion. Under CUT the physical area where the explosion occurred has only changed in respect to the character of the objects and energy occupying it; its size has not changed at all. Thus SST is right on the question of whether the universe has held "steady" and not expanded in the physical area in question, even if SST is right for the wrong reasons. (Contrary to SST, though, new matter was and is not being created from nothingness in a steady, ongoing, eternal process. Under CUT the matter that brought our universe into being was already there, in the form of a parent universe star that was located in the surrounding space of the total cosmos. That star exploded and its mass took on different form.)

So far as BBT is concerned relating to the expansion controversy with SST: Big Bangers are right that expansion is occurring, but wrong in their proffered reason for the expansion; wrong regarding the mechanics of the expansion, wrong regarding the thing that is expanding (a stretching of space between galaxies as opposed to things moving within the universe); wrong regarding the reason for expansion's increased rate of expansion; and wrong as to the entity into which the universe is expanding.

Expansion under CUT is much more reasonable. Under CUT there were actual pre-existing physical locations (1) where the parent universe progenitor star was situated before it exploded, and (2) in the surrounding space into which the remnants of the exploded progenitor star have moved and are presently moving. Therefore, there is only an "expansion" under CUT in the sense of things moving away from the beginning point of the explosion in the way things move away from any other blast in any other known place. The blast field following the SUSE expanded, but there has been no "expansion" in the nature of BBT's claimed (a) creation of new space, and (b) increase in the size of the total universe.

Back to whether there are other precepts of BBT that can be incorporated to some extent into CUT.

In CUT we can also preserve the idea that our universe started with some sort of a bang. Big Bangers refuse to acknowledge that it was an explosion, or that it was *matter* (that didn't explode), or that it occurred in any place, or that it occurred in time (whoops, here we go again), etc. But they do admit that there was an initial hot *something* that occurred. Good. Let's retain that. The cosmic background radiation we can measure today appears to confirm an original hot cataclysm.

Next BBT concept to salvage: The concept of nucleosynthesis of light elements seems to fit the bill, even though the causation of the beginning cauldron suggested by BBT is a world apart from the reality of a SUSE. But the production and/or distribution of light elements as a result of a beginning hot bang in our infant universe is something BBT and CUT can mostly agree upon.

Also, nothing in CUT says that we should abandon BBT's concept that our universe is evolving (from simple to more complex, from mostly lighter elements to heavier elements, etc.). BBT says that it's the infinitesimally small singularity that is evolving; CUT says that it's the remnants of the SUSE that are evolving; but let's retain the concept of generational star evolution within our universe.

CUT fully incorporates BBT's concepts of the (a) applicability of relativity and the (b) applicability of the universality of laws of physics all through our universe. In fact, as seen throughout this book, it's a basic premise of CUT that the laws of physics in our universe are merely an extension of the physical laws of our vast surrounding cosmos.

What about the age of our universe? There may well be at least some adjustments to BBT's calculation of the age of our universe under CUT once BBT's inflationary cheat factor is eliminated and the beginning size and nature of our SUSE are taken into account. BBT's singularity and CUT's SUSE are two entirely different animals, conceptually and physically.

•••••

I'm sure there are many other aspects of cosmology that will leap into reconsideration if Cosmic Unity is adopted as the standard model of cosmology. CUT gives us all that we need to begin an entirely fresh approach to, and analysis of, many cosmological questions we face today.

Obviously, many of the forgoing observations and questions relating to Cosmic Unity will not directly affect the lives of us earthly mortals and our descendants for many moons to come. But isn't it interesting to think about our world with the assurance and security that our universe had a comprehensible beginning, and is a cohesive part of a large, interconnected, unified, pre-existing cosmos? Think of all the answers that such knowledge provides, and all the possibilities and hope that such a total world offers for us.

Notes

1–Introduction
1. Alex Vilenkin, *Many Worlds In One,* Hill and Wang, New York, 2006, p. 5.
2. Kip S. Thorne, *Black Holes And Time Warps*, W.W. Norton & Company, Ltd. New York 1994, p. 558.

2–The Official Position
1. Timothy Ferris, *The Whole Shebang*, Simon & Schuster Paperbacks, New York 2005, pp. 14-17.
2. Paul Davies and J. Gribbin, *The Matter Myth*, Simon & Schuster/Touchstone, New York, 1992, p. 122.
3. Stephen W. Hawking, *A Brief History Of Time,* Bantam Books, New York, 1990, p. 46.
4. Question posed to Bob Hirshon from *Science Update (*from *Science NetLink)* athttp://www.sciencenetlinks.com/sci_update.cfm?DocID=7:

   "I want to know what happened before the big bang.

   Well, Phil, we spoke to Nobel Prize winning Physicist Leon Lederman, of the Illinois Institute of Technology. Lederman: 'Well, the first thing is there's no "before." Because time itself, as far as we understand time, was generated—and space—at the Big Bang.'"

5. Janna Levin, *How The Universe Got Its Spots*, Anchor Books/Random House, New York, 2003, p. 90.
6. Ibid., p. 94.
7. Simon Singh, *Big Bang*, Harper/Perennial, New York, 2004, p. 490.
8. Janna Levin, *How The Universe Got Its Spots*, p. 90.
9. NASA Goddard Space Flight Center Home Page "*Ask An Astrophysicist*" Q/A at http://imagine.gsfc.nasa.gov/docs/ask_astro/answers/060629a.html.
10. Brian Greene, *The Fabric of the Cosmos*, Vintage Press, New York, 2005, p. 240.
11. Paul Davies, *Cosmic Jackpot,* Houghton Mifflin Company, New York, 2007 p. 69.
12. *Curious About Astronomy? Ask An Astronomer* website, Cornell University, at http://curious.astro.cornell.edu/question.php?number=248.
13. Jennifer Ouellette, *The Physics of the Buffyverse*, Penguin Books, New York, 2006, p. 196.

14. Dennis Overbye, *What Was There Before the Big Bang? New York Times*, May 22, 2001 at http://www2.gol.com/users/ coynerhm/before_ the_big_ bang_there_was__.htm.
15. UCLA Professor Eugene L. (Ned) Wright at http://www.astro.ucla.edu/~wright/b4u-write.html:
    Everything that we measure is within the Universe, and we see no edge or boundary or center of expansion. Thus the Universe is not expanding into anything that we can see, and this is not a profitable thing to think about.
16. Neil De Grasse Tyson, *Death by Black Hole*, W.W. Norton & Company, Inc., New York, 2007, p. 345.
17. Among others see, for example, Simon Singh, *Big Bang*, p. 491, and Fred Adams, *Our Living Multiverse*, Pearson Education, Inc., New York, 2004, p. 40.
18. Alex Vilenkin, *Many Worlds In One*, p. 12.
19. ....The other being whether intelligent life exists in the universe outside Earth. Scientists would probably add another—which is maybe less metaphysical, but more scientifically compelling—and that relates to whether there is an all-embracing "theory of everything."
20. We are said to need about 5 hydrogen atoms for every cubic meter in the universe to reach critical density. See for example, Brian Greene's books *The Elegant Universe,* Vintage Books/Random House, Inc., New York, 2003, p.234 and *Fabric of the Cosmos*, p. 242.

3–The Simplest Solution: Cosmic Unity Theory
1. I'll use the terms "larger universe," "surrounding universe," total universe," "parent universe" and similar phrases interchangeably. As I'll mention later, (Note 4) there are a few references by others to various types of cosmologically-related universes that are totally different from anything I shall be referring to. Because of such differences I'm reluctant to continually talk in terms of a parent universe. I also hesitate to get too anthropomorphic about things, even though I think we're related to the large, surrounding universe in the first degree of consanguinity. So, when I talk in terms of a parent universe I'm simply referring to the surrounding cosmos, or medium, of which we form a part.
2. Some say that our universe is presently as much as 92 to 94 billion light years in diameter. The graphics illustrator assumed for the purpose of his project that the observable universe of about 14 billion light years equals the real co-moving diameter of our universe as it exists today. But the point remains the same. See Note 7 of Chapter 4.
3. We must keep in mind, of course, that today our universe's farthest point from us is much more distant than 14 billion light years, because today

we're seeing the light that began its journey toward us 14 billion years ago. Our universe has expanded greatly in all directions since then.

4. There are at least a couple cosmological theories that mention some sort of a "parent universe"; however I can't find anything close to the type of parent I'm talking about. For example, Dr. Rag Baldev mentions a parent universe, but it is nothing like what we're discussing. His involves a first Big Bang that preceded a second Big Bang, with a delay of about one billion years. He speaks in terms of an Outer Reservoir and a Central Reservoir, and talks about two Primitive Black Holes merging, then recycling and rejecting material. He further thinks of the universe as composed of some "seven circuits" in infinite space where the galaxies are evidently constantly making adjustments in their circuits, but the universe itself is not really expanding.

Another "parent" candidate (even if not specifically identified as a parent) is a black hole itself. Under some black hole theory our universe is somehow located within a black hole. In one such theory new matter is constantly pouring into it at a speed faster than the speed of light; and the matter entering our universe is actually "dark energy." Black hole concepts don't necessarily refer to "parent" universes, but do suggest possibilities of different physics and geometries in each such possible black-hole-confined universe. Hawking's hypothetical "baby universes" that could emerge from black holes in "imaginary time" imply contemplation of a sort of parent universe related to a black hole.

Another theory talks of a child universe created from a "false vacuum bubble." The child universe somehow detaches from a "true vacuum" and then continues on with its life completely disconnected from what could be considered the parent.

Then there is a concept of Andre Linde's new universes "self-creating" from tiny bubbles from a parent universe's foam, which self-created bubble universes in turn sprout other bubble universes and wormholes. The bubble universe concept involves creation of universes from the quantum foam of a "parent universe." On very small scales, the foam is frothing due to energy fluctuations. It is these fluctuations that may create tiny bubbles and wormholes. If the energy fluctuation is not very large, a tiny bubble universe may form, experience some expansion like an inflating balloon, and then contract and disappear from existence. However, if the energy fluctuation is greater than a particular critical value, a tiny bubble universe forms from the parent universe, experiences long-term expansion, and allows matter and large-scale galactic structures to form.

The "self-creating" aspect of Andre Linde's self-creating universe theory stems from the concept that each bubble or inflationary universe will sprout other bubble universes, which in turn, sprout more bubble universes. Linde feels that the universe we live in has a set of physical constants that seem tailor-made for the evolution of living things.

Martin Rees makes a general reference to a "daughter" universe, but, as noted earlier, believes that if such a daughter relationship could ever exist "no information could be exchanged with a daughter universe," even though such daughter could possibly bear the imprint of its parentage. See Martin Rees, *Before the Beginning*, Helix Books/Perseus Books, United States 1997, p. 4.

There is also a great variety of other multiverse, megaverse, pocket universe, parallel universe and alternate universe theories. And I'm not even mentioning here the multifarious "other dimension" concepts, string theory and non-string theory, supergravity, M-theory, wormholes, branes, etc. And who among us has not considered the possibility that our universe is really just a microscopic particle of a larger particle of a larger one, etc.?

There could be some truth in one or more of these theories. But for now let's look at the simplest and most reasonable explanation for our universe coming into being.

5. Whatever the actual number is to reach the farthest extent of our universe, I'm talking about moving out past that distance, thus exiting our universe and immediately entering the realm of the surrounding universe of which we form an integral part. (As we know, the farthest boundaries of our universe have expanded outward considerably since the light that we see began its journey about 14 billion years ago.)
6. A good discussion of the Standard Model is found in Michio Kaku's *Hyperspace*, Anchor Books/Random House, Inc., New York, 1995, pp. 121-125.
7. I realize that some scientists like to say that our universe has no "boundaries" or "edges" or "sides" in the normal sense of the words (which I believe are further accommodations to BBT), but such references serve our purpose here insofar as the word "boundaries" applies here to the physical limits of our universe's existence. I'll discuss the issue of boundaries for our universe later in this book.
8. See, for example, the cover of the Fred Adams book, *Our Living Multiverse,* PI Press, New York, 2004; and *Hyperspace*, Kaku, pp. 19-20.
9. Adams, *Our Living Multiverse* p. 211.
10. Ibid., p. 213.
11. Michio Kaku speaking, at BBC.co.uk *Science and Nature, Radio & TV Follow-up*, BBC Two 9:00 pm., Thursday, February 14, 2002, Narrator Dilly Barlow.

12. Michael Duff speaking, Ibid.
13. Singh, *Big Bang*, p. 503.
14. Alex Vilenkin, *Many Worlds In One*, p. 204.
15. Brian Greene, *The Elegant Universe*, p. 366.
16. Ibid., p. 367.
17. Martin Rees, *Before the Beginning*, p. 3.
18. Id.
19. Ibid., p. 7.
20. Ibid., p. 249.
21. This was an answer by Rees to a question regarding the multiverse noted at www.edge.org.
22. In his Stanford University website (http://www.stanford.edu/~alinde/) Mr. Linde refers to what he conceives to be the differing laws of physics and differing cosmological constants in eternally self-generating universes, as follows:

    > After inflation the universe becomes divided into different exponentially large domains inside which properties of elementary particles and even dimension of space-time may be different. Thus the universe looks like a multiverse consisting of many universes with different laws of low-energy physics operating in each of them.
    >
    > ....
    >
    > The idea of an inflationary multiverse (the universe consisting of many universes with different properties) was first proposed in 1982 in my Cambridge University preprint *Nonsingular Regenerating Inflationary Universe*.
    >
    > ....
    >
    > The main goal of these two papers was to propose a physical mechanism which would allow the existence of different exponentially large parts of the universe with different values of the cosmological constant.
    >
    > ....
    >
    > This means that our multiverse may consist of exponentially many exponentially large domains (universes), each of which may live in accordance to one of the exponentially large variety of laws of the low-energy physics.

23. Quoted from Editor Andrew Chaikin, *Are There Other Universes? Space & Science*, posted February 5, 2002 at 7:00 AM. See http://www.space.com/scienceastronomy/generalscience/5mysteries_universes_020205-1.html.
24. Id.
25. Paul Davies, *Cosmic Jackpot*, p. 33.

26. Ibid., p. 176.
27. Lisa Randall, *Warped Passages*, Harper/Perennial, New York, 2005, p. 61.
28. Max Tegmark has categorized multiverse possibilities into what he conceives of as four levels, but from what I can see none of the levels would be able to correspond with each other, for various reasons, including differing physical constants, differing fundamental equations of physics, and differing "states" of being, etc. See *Wikipedia* Multiverse. http://en.wikipedia.org/wiki/Multiverse_(science).
29. Paul Davies, *Cosmic Jackpot*, p. 172.
30. *COSMOS: Before There Was Light*, Astronomy Collector's Edition 2006, Kalmbach Publishing Company, p. 34.
31. Ibid., p. 37.
32. William of Ockham (Occam) (1284?-1387?) was a leading English theologian and professor who stressed that in science the simplest theory that fits the facts of a problem should be the one selected. This principle became known as "Ockham's Razor." *World Book Encyclopedia*, Vol. 21, p. 260.
33. There is a controversy about whether there was ever a true argument about such a hypothesis. Scholars can't seem to find a basis for the oft-stated inquiry. Aquinas evidently did face up to an inquiry regarding whether several angels could be in the same place at once (to which he answered no). See *A Straight Dope Classic from Cecil's storehouse of human knowledge*—http://www.straightdope.com/ classics/a4_132.html.
34. Paul Davies points out (as do others) that under the second law of thermodynamics any irreversible processes in the universe that proceed at a finite rate (ie., the collapse of a star) will reach their finite stage in a finite period of time. Like a clock that is running down, the Great Cosmic Clock should have stopped clicking by now. *Cosmic Jackpot*, p. 71.
35. Stephen Hawking, *Black Holes and Baby Universes and Other Essays*, Bantam Books, 1994, pp. 121-125.

4–Problems For Big Bang No Problem For Cosmic Unity

1. Let's not haggle over the word "proofs." Sometimes they're referred to as "proofs for" BBT, sometimes "evidence of" BBT, sometimes "pillars supporting" BBT, etc.
2. Such *"Open Letter to the Scientific Community"* was published in the *New Scientist*, May 22, 2004.
3. See Lawrence M. Krauss, *ATOM An Odyssey from the Big Bang to Life on Earth...and Beyond*, Little, Brown & Company, New York, 2001, p. 15.
4. Timothy Ferris, *The Whole Shebang*, p. 70
5. Some scientists believe that the universe has too much large scale struc-

ture in the form of various interspersed voids and walls to have developed form in so short a time as only about 15 billion years. At the average speeds of galaxies' expansions the giant superclusters would need many more times 15 billion years to form—and that is even assuming that the galaxies were "direction oriented," which they weren't. They presumably were built up by gravitational action (a slower process) alone.

6. See reference by Brian Greene in *The Fabric of the Cosmos,* p. 285:

> Roughly, 10^{-35} seconds after the burst began, the inflation field found its way off the high-energy plateau and its value throughout space slid down to the bottom of the bowl, turning off the repulsive push.

7. From the official *String Theory* website:

> The cosmic microwave background is the cooled remains of the radiation density from the radiation-dominated phase of the Big Bang. Observations of the cosmic microwave background show that it is amazingly smooth in all directions, in other words, it is highly isotropic thermal radiation. The temperature of this thermal radiation is 2.73° Kelvin. The variations observed in this temperature across the night sky are very tiny.
>
> Radiation can only be so uniform if the photons have been mixed around a lot, or thermalized, through particle collisions. However, this presents a problem for the Big Bang model. Particle collisions cannot move information faster than the speed of light. But in the expanding Universe that we appear to live in, photons moving at the speed of light cannot get from one side of the Universe to the other in time to account for this observed isotropy in the thermal radiation. The horizon size represents the distance a photon can travel as the Universe expands. The horizon size of our Universe today is too small for the isotropy in the cosmic microwave background to have evolved naturally by thermalization. So that's the horizon problem. *[http://www.superstringtheory.com/cosmo/cosmo4.html]*

8. Brian Greene, *The Fabric of the Cosmos,* p. 290.
9. See *"Open Letter to Scientific Community," New Scientist* magazine, May 22, 2004.
10. See, for example, Alec MacAndrew's *"The Big Bang Is Not A Myth"* at http://www.evolutionpages.com/big_bang_no_myth.htm:

> The fact is that homogeneity was a real problem for Big Bang and one that has been resolved....the Horizon Problem (and the Flatness Problem) has been resolved by the concept of inflation...

11. Id.

12. See The Stanford University website http://www.stanford.edu/~alinde/.
13. Timothy Ferris states that the rule that nothing can be accelerated faster than the speed of light "does not apply to galaxies in an expanding universe. That rule is true in static space, but expanding cosmic space can carry galaxies away from one another at velocities greater than that of light." Timothy Ferris, *The Whole Shebang*, p.44. The same general explanation for faster-than-light inflationary expansion is given by others. See, for example, Brian Greene's *Fabric Of The Cosmos*, p. 523, Note 16.
14. "Physicist Stephen Hawking and mathematician Roger Penrose in 1970 are said to have conditioned their proof of BBT on both the correctness of Einstein's theory of relativity and on there only being so much matter in the universe as we can observe. Even though the latter is probably untrue, for now the BBT still reigns supreme." *INTRO TO ASTRONOMY*, http://earthguide.ucsd.edu/virtualmuseum/ ita/03_1.shtml.
15. See *Cosmic Jackpot*, Paul Davies, p. 141: "The nuclear material of the star collapses to a density of almost a billion tons per cubic centimeter—so dense that even neutrinos have a tough job plowing through it." [1 billion x 2,000 lbs. = 2 trillion lbs.]
16. The cosmological constant refers to a theoretical force that was introduced by Einstein. The cosmological constant was to counteract gravity so as to result in the constant, static, eternal universe conceived of by him. He later retracted the theory (after discovery of the Hubble redshift), but today the cosmological constant is being reintroduced by some cosmologists as a theoretical "dark" force that can be either attractive and repulsive, depending on various circumstances.
17. Lisa Randall, *Warped Passages*, pp. 298-299.
18. The inverse square law of gravity holds that the force of gravity diminishes with the distance as the *square* of the separation between bodies.
19. In the 1990s two separate groups of astronomers determined that the rate of expansion of our universe was accelerating, not decelerating, as had been anticipated. They calculated that our universe had been expanding at an accelerated rate since it was about 7 billion years old (now it's thought to be about 14 billion years old). Thus it could well be (my theory, not theirs) that the material in our universe has been noticeably interacting with structures in our next larger universe—thus accelerating in expansion—for at least half the lifetime of our universe.
20. See for one of many such statements Brian Green, *The Elegant Universe*, p. 83.

5–Big Bang Evidence Proves Cosmic Unity Theory
1. Although the smoothness of our universe has been used as an argument for BBT some have turned the "smoothness" argument against BBT. Some anti-BBT scientists contend that various locations in our universe that are

farther apart than light can have traveled since our universe began can have no effect on each other. Thus the farthest-apart celestial bodies don't have a logical explanation under BBT.

2. For example, Meta Research (a decidedly anti-BBT site) states this regarding the CMB proof for the Big Bang at http://metaresearch.org/cosmology/BB-top-30.asp:

> The microwave "background" makes more sense as the limiting temperature of space heated by starlight than as the remnant of a fireball. The expression "the temperature of space" is the title of chapter 13 of Sir Arthur Eddington's famous 1926 work, Eddington calculated the minimum temperature any body in space would cool to, given that it is immersed in the radiation of distant starlight. With no adjustable parameters, he obtained 3°K (later refined to 2.8°K), essentially the same as the observed, so-called "background," temperature. A similar calculation, although with less certain accuracy, applies to the limiting temperature of intergalactic space because of the radiation of galaxy light. So the intergalactic matter is like a "fog," and would therefore provide a simpler explanation for the microwave radiation, including its blackbody-shaped spectrum.
>
> Such a fog also explains the otherwise troublesome ratio of infrared to radio intensities of radio galaxies.

3. This temperature is about 100 times higher than that created by a large-mass star supernova in our universe. This, of course, may still be too low; although sundry other considerations enter in. See *Many Worlds In One*, by Alex Vilenkin, p, 37 regarding temperatures in large star core collapses. For another reference to temperature of 1 trillion degrees a microsecond after the big bang see Paul Davies, *Cosmic Jackpot*, p. 142.
4. See for these density, temperature and size calculations *COSMOS,* by John Gribbin and Simon Goodwin, Magpie Books, a Constable & Robinson Ltd. imprint, London, 2006, p. 24.
5. Of course no elements were produced in the immediate aftermath of the Big Bang. It was much too hot for anything other than an intense plasma soup to exist. The light elements only formed after the hot plasma cooled to the point that electrons could began attaching to the nuclei of atoms.
6. See *Astronomy & Astrophysics Standard Cosmic Ray Energetics and Light Element Production*, by B. D. Fields, K. A. Olive, M. Cass, and E. Vangioni-Flam.
7. For reference to a discussion of a hypothetical negative pressure resulting in a hypothetical negative gravity one can see Brian Greene's *The Fabric Of The Cosmos*, pp. 273-279.

6–What Kind Of SUSE Brought Us Into Being?
1. The anthropic principle holds that if things weren't structured in just the way they are on our planet, in our solar system, in our galaxy, etc. we wouldn't be here at all. We're here because if conditions were different we wouldn't be here. In other words, there wouldn't even be anyone around to ask the question or know how to ask a question about who we are and why we're here or to wonder what things would be like if conditions in the universe were different.
2. I even wonder if one can one rule out that what remained in place after our SUSE was a pulsar?
3. Timothy Ferris, *The Whole Shebang*, p. 18.

7–More About the Physical Connection Between Our Universe and the Surrounding Universe, and Random Thoughts
1. Einstein originally felt that our universe is static, and in fact came up with a theory of dark energy, later referred to as a type of "cosmological constant," to counterbalance what would be the apparent tendency of our universe to crush inward and die of its own weight if in a static state. He later abandoned such early concept of a static universe after Edwin Hubble in 1927 provided observational evidence that our universe was expanding. As previously noted, Big Bangers have now resurrected the concept of a cosmological constant in order to explain why our universe is accelerating its rate of expansion—a phenomenon otherwise unexplained under BBT.
2. A good discussion of scale is found in *The View From the Center of the Universe,* by Joel R. Primack and Nancy Ellen Abrams, Riverhead Books, New York, 2006, pp.166-73.
3. Peter Ward, in *Life As We Do Not Know It*, Viking/Penguin Group, New York, 2005, devotes an entire book to the chemical, physical, and biological components that might apply to alien life forms in just our solar system and our universe alone.

Index

A

accelerated expansion, See universe, expansion of,
Adams, Fred, 33, 115 n17, 117 n8, 117 n9
Andromeda, 98
anthropic principle, 39, 89, 123 n6-1
Aquinas, St. Thomas, 26, 119 n33
Aristotle, 39

B

Baldev, Rag, 116
balloon analogy, 53, 79, 116 n4
Big Bang Theory (BBT), 7-11, 12-21, 23-26, 28, 29, 37, 38, 39, 41, 44, 45, 46-73, 74-88, 94, 109
  arguments, proofs for, 74-88
  cyclic, 20, 21, 48, 61, 63, 70, 71, 87
  developed structure after, 54, 55, 59, 64, 83, 120 n5
  elements, light elements production, 84-88, 100-101, 112-113
  expansion, See universe, expansion
  horizon problem, 30, 38, 55-58, 64, 87, 106, 111, 120 n7, 121 n10
  inflation, 19, 49, 55-60, 64, 81-83, 85, 88, 113, 120 n6, 121 n10, 121 n13
  origination from nothing, 12, 14-20, 23, 47-50, 54, 55, 59, 62, 69, 85, 94, 110
  problems with, origination, time, beginning, expansion under, 46-73
  Standard Model, 10, 13-14, 20, 36, 45, 49, 62, 117 n6
  time began with, 13-16, 18, 23, 44, 47, 49-50, 59, 69, 80, 82, 85, 110, 112, 114 n4
  time-related problems, 49, 50, 54-56, 59, 60, 69, 71, 83, 110, 120 n5, 120 n7
  trace-back error, 23, 62-63, 65, 79-81, 111
  Vatican, 17
Big Crunch, 20, 49, 61-63, 70, 72, 86-87, 96, 106
Big Rip, 65, 72, 100
Burbidge, Geoffrey, 84
Burbidge, Margaret, 84
blackbody radiation, 74, 122 n2
black holes, 27, 31, 44-45, 58, 61, 87, 92, 94, 99-100, 116 n4,
branes, 36, 40, 42, 106, 117 n4
bubble universe(s), 31, 32, 35, 36, 40, 42, 94, 106-107, 116 n4, 117 n4

C

Chaikin, Andrew, 36, 118 n23
CIZA, Clusters In the Zone of Avoidance, 68
COBE, See Cosmic Background Explorer
constants of nature, 34, 35, 37, 108, 117 n4, 118 n23, 119 n28
Cornell University, 16, 114
Cosmic Background Explorer, 75
cosmic background radiation (CMB), 12, 55, 56, 57, 74-78, 79, 88, 112, 120 n7, 122 n2
cosmological constant, 37, 63-64, 66, 79, 110, 118 n22, 121 n16, 123 n1

124

cosmological principle, 13
Crab Nebula, 76, 90
Curtis, Heber D., 30

D

dark energy, 40, 64-67, 70, 109-110, 116 n4, 121 n16, 123 n7-1
dark matter, 21, 70
Davies, Paul, 14, 16, 36, 37, 38, 119 n34, 121 n15, 122 n3
Duff, Michael, 33

E

Eddington, Arthur, 53, 122 n2
Einstein, Albert, 13, 26, 30, 52, 59, 64, 66, 78, 79, 102, 121 n14, 123 n7-1
entropy, 27-28
expansion, See universe, expansion

F

Ferris, Timothy, 52, 93, 121 n13
Fowler, William, 84

G

Great Attractor, 68
Greene, Brian, 34, 52, 56, 115 n20, 120 n6, 121 n13, 123 n7
Guth, Alan, 16-17, 19, 55

H

Haldane, J.B.S., 53
Hawking, Stephen, 8, 14, 45, 61, 116 n4, 121 n14
heat death, 27
heavy elements, 84-85, 88, 90, 100-101
helium, 9, 84-85, 89
Helmholtz, Hermann von, 27-28

homogeneous, homogeneity, 13, 57, 76, 121 n10
horizon, 9-10, 25-26, 29-30, 38, 48, 55-58, 64, 67, 72, 77, 83, 87, 92-93, 106, 111, 120 n7, 121 n10
Hoyle, Fred, 84
Hubble, Edwin, 79, 121 n16, 123 n7-1
hydrogen, 9, 61, 84-85, 89, 115 n20

I

infinite, infinity, 31, 33-40, 43, 58, 65, 104, 106, 116 n4
inflation, See Big Bang Theory, inflation
isotropic, isotropy, 13, 74, 76, 78, 120, n7

K

Kaku, Michio, 33, 117 n6, 117 n7, 118, n11

L

Layzer, David, 28
Lederman, Leon, 14, 114 n4
Lemaitre, Georges, 78-79
Levin, Janna, 15, 52
light elements, 84-88, 91, 100-101, 112-113, 122 n5
Linde, Andrei, 16, 17, 35, 116 n4, 117 n4, 118 n22, 121 n12

M

megaverse, See multiverse generally
Milky Way Galaxy, 14, 30, 68, 76, 82, 92, 98, 102, 106
missing matter, 70-71
molecular gas clouds, 90-91
multiverse theory generally, 11, 31-

39, 72, 110, 117 n4, 117 n8, 118 n21, 118 n22, 119 n28
CUT differs from, 105-107

N

NASA, 15, 75
Newton, Isaac, 26, 32
nothingness, nonexistence, 12, 14, 18, 23, 25, 42, 44, 48, 50-55, 59, 62, 69, 72, 79-80, 83, 85, 112
nucleosynthesis, 56, 84, 100, 112

O

observable universe, See universe, visible,
Ockham, William of, 39, 109, 119 n32
Ockham's Razor, 39
Open Letter to Scientific Community, 56, 120 n9
open universe, See universe, open
Overbye, New York Times, 16, 17

P

parent universe, 22-23, 25-26, 28, 30-33, 36-38, 40-47, 49-50, 53-54, 67-68, 71-72, 75, 77, 81, 85-87, 90-105, 107-110, 112, 115 n1-117 n4
 illustration, 95
Penzias, Arno, 74, 79
phase transitions, 108
Pius XII, 17
Planck, Max, 26, 28, 111
planetary nebula, 67, 91-93
pocket universe(s), 32, 36, 39, 106, 107
Primack, Joel, 104, 123 n2
progenitor star, 23, 37, 41-42, 46-47, 58, 85-86, 90-91, 93, 96, 100, 108, 112
Puppis, 90

Q

quantum fluctuation, 48
quantum foam 35, 116 n4
quantum physics, 18, 23, 26, 47-48, 111

R

Randall, Lisa, 36, 64
redshift, 79, 111, 121 n16
Rees, Martin, 35, 117 n4
relativity, 13, 52, 65, 113, 121 n14

S

Sagan, Carl, 26
scale considerations, 32, 42-43, 67, 76, 91, 95, 97, 104, 107-109, 123 n2
Shapley, Harlow, 30
Shapley Supercluster, 68
Singh, Simon, 15, 34, 114 n17
singularity, 13, 15, 18, 29, 44, 46, 49-51, 58-63, 65, 72, 75, 78-81, 83-84, 86-87, 100, 110, 111, 113
Smolin, Lee, 8, 44
solar system, 14, 24, 37, 44, 47, 66, 85, 90, 98, 102, 123 n6-1, 123 n7-3
space, spacetime, 10, 12-15, 23-25, 31, 36, 42, 44-45, 49-57, 59-60, 62, 64-66, 70, 76-77, 82, 87, 91, 94-95, 107, 111, 112, 114, 116 n4, 118 n22, 120 n6, 121 n13, 122 n2
static universe, 102, 111, 121 n13, 121 n16, 123 n7-1
Steady State Theory (SST), 12, 13,

48, 79, 102, 103, 111-112
Steinhardt, Paul, 35-36
supernovae, 31-32, 37, 38, 41-43, 45, 47, 51, 61-62, 65, 67, 70, 75-77, 80-81, 84-87, 89- 93, 95, 100-101, 104-108, 122 n3
Surrounding Universe Star Explosion (SUSE), 23, 31, 41-49, 51, 53, 55, 58-60, 62-63, 66-67, 70-71, 75-79, 81-89, 91-96, 98-100, 102-104, 112-113, 123 n6-2,
symmetry, 26, 37

T

Tegmark, Max, 38, 119 n28
Thorne, Kip, 10
time, 10, 13-16, 18, 23, 25, 27, 30-31, 34, 41-42, 44-45, 47-50, 54-56, 59-60, 62, 66, 69, 71, 75, 78, 80, 82-83, 85, 92, 98, 103, 110, 112, 114 n4, 116 n4, 118 n22, 119 n34, 120 n5, 120 n7
trace-back error, See Big Bang Theory, trace-back error
Tyson, Neil De Grasse, 17

U

universality principle, 13, 37, 113
universe, our, definitions, 9-13
   age, 9, 33, 54, 64, 82-85, 88, 106, 113, 121 n19
   boundaries or edges, 15, 29, 50-52, 54, 60, 64, 68, 77, 92, 96, 110, 115 n15, 117 n5, 117 n7
   closed, 27-29, 49, 63, 71-72, 106
   critical density, 20, 63, 115 n20
   curvature of, 52, 66
   evolution of, 12-13, 23-24, 37, 42-43, 45-47, 57, 62, 82, 86, 88-90, 94-95, 98, 100-101, 104, 108, 113, 117 n4, 120 n7

   expansion of, 12-13, 19-20, 23, 25, 31, 35, 41, 43, 45-46, 49-57, 59-72, 75-76, 78-83, 85, 88, 93-98, 100, 102-104, 106, 110-112, 115 n15, 116 n3, 116 n4, 117 n5, 120 n5, 120n7, 121n13, 121 n19, 121 n7-1
   location, structure of, 7, 11, 23, 25, 31-32, 36-38, 41-49, 51, 53, 55, 60, 67, 76-77, 81, 89-91, 94-99, 104-105, 110-113
   non-verticality of, 43, 104
   observable, See universe, visible,
   open, 28, 49, 63, 72, 98, 106, 110
   order in, 26-29, 41, 66
   smoothness of, 55, 58, 64, 76-77, 82, 121 n1
   visible, 9-11, 13, 24-25, 29-30, 32, 37, 42, 47-48, 57, 67-68, 77, 82, 87, 92-93, 95, 115 n2

V

vacuum energy, 47, 64-66, 110
Vatican, 17
Veil Nebula, 90
Vela, 90
Vilenkin, Alex, 10, 19, 34, 38, 122 n3
Virgo Supercluster, 24, 68
visible universe, See universe, visible,

W

Ward, Peter, 123 n3
Wilkinson Microwave Anisotropy Probe (WMAP), 52, 75
Wilson, Robert, 74, 79
WMAP See Wilkinson Microwave Anisotropy Probe
wormhole(s), 31, 35, 116 n4, 117 n4
Wright, Ned, 52, 115 n15